AF614706

METHODS IN MOLECULAR BIOLOGY™

Series Editor
John M. Walker
School of Life Sciences
University of Hertfordshire
Hatfield, Hertfordshire, AL10 9AB, UK

For further volumes:
http://www.springer.com/series/7651

Vascular Proteomics

Methods and Protocols

Edited by

Fernando Vivanco

Department of Immunology, IIS-Fundacion Jimenez Diaz, Madrid, Spain;
Department of Biochemistry and Molecular Biology I, Universidad Complutense, Madrid, Spain

Editor
Fernando Vivanco
Department of Immunology
IIS-Fundacion Jimenez Diaz
Madrid, Spain

Department of Biochemistry and Molecular Biology I
Universidad Complutense
Madrid, Spain

ISSN 1064-3745 ISSN 1940-6029 (electronic)
ISBN 978-1-62703-404-3 ISBN 978-1-62703-405-0 (eBook)
DOI 10.1007/978-1-62703-405-0
Springer New York Heidelberg Dordrecht London

Library of Congress Control Number: 2013934715

Printed on acid-free paper

Humana Press is a brand of Springer
Springer is part of Springer Science+Business Media (www.springer.com)

Preface

Proteomics is a rapidly expanding investigation platform in cardiovascular medicine. Driven by major improvements in Mass Spectrometry (MS) instrumentation and data analysis, the proteomics field has flourished in recent years particularly in the study of complex diseases. These recent advances are characterized by the development of quantitative MS-based methods that moved the field on from primarily identifying proteins to also providing measurements of relative changes in protein levels between different cell states, typically normal controls versus diseased samples. The application of proteomic techniques to the vascular pathology is a true reflection of this progress. Vascular proteomics of atherosclerotic lesions has become a major experimental group of tools in the last years notably by revealing novel proteins and signaling pathways to the processes involved in atherogenesis and its complications. This book encompasses a selection of techniques and methods that target the key processes implicated in atheroma plaque composition, formation, and development and the various components and plasma proteins involved in the atherosclerotic process. In this volume dedicated to Vascular Proteomics, we provide protocols and up-to-date methods for the analysis of arteries, cells, lipoproteins, body fluids (plasma, urine), and metabolites, with a particular focus on MS-based methods of protein and peptide quantification. The development of LC-MS/MS approaches has been a significant advance in many areas of biomedical research, and vascular proteomics is no exception. After an introductory chapter, we have followed a hierarchical order in the different chapters starting with methods dedicated to tissues (Laser microdissection for the analysis of human arteries; Use of SIMS-TOF in vascular Biology), cells (Vascular smooth cells treated with atenolol; Phosphoproteomic analysis of aortic endothelial cells; Characterization of membrane and cytosolic proteins of erythrocytes), lipoproteins (Quantitative analysis of apolipoproteins in HDL by top-down differential mass spectrometry; Quantitative proteomic analysis of high-density lipoproteins by stable 18-O-isotope labeling), plasma and plasma proteins, and metabolites (Biomarkers of abdominal aortic aneurism by iTRAQ analysis of depleted plasma; Absolute Quantitation of proteins in human blood by multiplexed multiple reaction monitoring mass spectrometry; MRM of plasma proteins in cardiovascular proteomics; Metabolites secreted by human atherothrombotic aneurysm). We have also included several chapters on very specific samples such as the analysis of thrombus (Identification of novel biomarkers of abdominal aortic aneurysm from thrombus conditioned media), the secretome of arteries (Characterization of human arterial tissue secretome by 2-DE and nLC-MS/MS), and the study of exosomes (Proteomic analysis of urinary exosomes in cardiovascular and associated kidney diseases by 2-DE and LC-MS/MS). This type of sampling has great potential for the possible identification of new biomarkers, and their analysis is therefore of great interest.

This book does not attempt to describe exhaustively all the techniques used in the field of vascular proteomics, rather is a representative selection of methods that can be a useful resource for experienced proteomics practitioners and specially for newcomers, in order to become acquainted with the practice of a selective group of proteomic techniques for cardiovascular research. The editors are especially grateful to all contributing authors for the time and effort they have put into writing their chapters, and particularly to the Methods in Molecular Biology series editor, John Walker, for his continuous advice and support through the editorial process.

Madrid, Spain *Fernando Vivanco*

Contents

Contributors

GLORIA ALVAREZ-LLAMAS • *Department of Immunology, IIS-Fundacion Jimenez Diaz, Madrid, Spain*

CORAL BARBAS • *CEMBIO (Center for Metabolomics and Bioanalysis), Facultad de Farmacia, Campus Monteprincipe, Universidad CEU San Pablo, Madrid, Spain*

MARIA G. BARDERAS • *Department of Vascular Physiopathology, SESCAM, Hospital Nacional de Parapléjicos, Toledo, Spain*

JUDITH A. BERLINER • *Departments of Pathology and Medicine, David Geffen School of Medicine, University of California, Los Angeles, CA, USA*

LUIS M. BLANCO-COLIO • *Vascular Research Lab IIS-Fundación Jiménez Díaz, Instituto de Investigacion Sanitaria Fundacion Jimenez Diaz (IIS-FJD), Universidad Autónoma de Madrid, Madrid, Spain*

CHRISTOPH H. BORCHERS • *University of Victoria - Genome British Columbia Proteomics Centre, University of Victoria, Victoria, BC, Canada; Department of Biochemistry and Microbiology, University of Victoria, Victoria, BC, Canada*

ELENA BURILLO • *Vascular Research Lab IIS-Fundación Jiménez Díaz, Instituto de Investigacion Sanitaria Fundacion Jimenez Diaz (IIS-FJD), Universidad Autónoma de Madrid, Madrid, Spain*

ENRIQUE CALVO • *Unidad de Proteómica, Centro Nacional de Investigaciones Cardiovasculares, CNIC, Madrid, Spain*

EMILIO CAMAFEITA • *Unidad de Proteómica, Centro Nacional de Investigaciones Cardiovasculares, CNIC, Madrid, Spain*

HELENE L. CARDASIS • *Proteomics, Merck and Co., Rahway, NJ, USA*

MICHAL CIBOROWSKI • *Centre for Clinical Research, Medical University of Bialystok, Bialystok, Poland*

ANDREW G. CHAMBERS • *University of Victoria - Genome British Columbia Proteomics Centre, University of Victoria, Victoria, BC, Canada*

LIWEI CHEN • *School of Chemical and Biomedical Engineering, College of Engineering, Nanyang Technological University, Singapore, Singapore*

WEI NING CHEN • *School of Chemical and Biomedical Engineering, College of Engineering, Nanyang Technological University, Singapore, Singapore*

VERÓNICA M. DARDÉ • *Proteomics Unit, SESCAM, Hospital Nacional de Paraplejicos, Toledo, Spain*

FERNANDO DE LA CUESTA • *Department of Vascular Physiopathology, Hospital Nacional de Parapléjicos, SESCAM, Toledo, Spain*

JESÚS EGIDO • *Vascular Research Lab IIS-Fundación Jiménez Díaz, Instituto de Investigacion Sanitaria Fundacion Jimenez Diaz (IIS-FJD), Universidad Autónoma de Madrid, Madrid, Spain*

HUIXING FENG • *School of Chemical and Biomedical Engineering, College of Engineering, Nanyang Technological University, Singapore, Singapore*

THOMAS G. GRAEBER • *Crump Institute for Molecular Imaging, Institute for Molecular Medicine, and California NanoSystems Institute, Jonnson Comprehensive Cancer Center, David Geffen School of Medicine, University of California, Los Angeles, CA, USA*

INMACULADA JORGE • *Laboratorio de Proteómica Cardiovascular, Centro Nacional de Investigaciones Cardiovasculares, Madrid, Spain*

BENEDICT JIA HONG LEE • *School of Chemical and Biomedical Engineering, College of Engineering, Nanyang Technological University, Singapore, Singapore*

XIANG LI • *School of Chemical and Biomedical Engineering, College of Engineering, Nanyang Technological University, Singapore, Singapore*

JUAN ANTONIO LOPEZ • *Unidad de Proteómica, Centro Nacional de Investigaciones Cardiovasculares, CNIC, Madrid, Spain*

AROA S. MAROTO • *Department of Immunology, IIS-Fundacion Jimenez Diaz, Madrid, Spain*

ROXANA MARTINEZ-PINNA • *Vascular Research Lab IIS-Fundación Jiménez Díaz, Instituto de Investigacion Sanitaria Fundacion Jimenez Diaz (IIS-FJD), Universidad Autónoma de Madrid, Madrid, Spain*

JOSÉ LUIS MARTIN-VENTURA • *Vascular Research Lab IIS-Fundación Jiménez Díaz, Instituto de Investigacion Sanitaria Fundacion Jimenez Diaz (IIS-FJD), Universidad Autónoma de Madrid, Madrid, Spain*

SEBASTIAN MAS • *Vascular Research Lab IIS-Fundación Jiménez Díaz, Instituto de Investigacion Sanitaria Fundacion Jimenez Diaz (IIS-FJD), Universidad Autónoma de Madrid, Madrid, Spain*

MATTHEW T. MAZUR • *Merck Research Laboratories, Rahway, NJ, USA; BioAnalytical Sciences, Imclone Systems (a wholly owned subsidiary of Eli Lilly and Co.), Branchburg, NJ, USA*

OLIVIER MEILHAC • *INSERM, U698, AP-HP, Hôpital Bichat, University Paris Diderot, Sorbonne Paris Cité, Paris, France*

JEAN BAPTISTE MICHEL • *INSERM, U698, AP-HP, Hôpital Bichat, University Paris Diderot, Sorbonne Paris Cité, Paris, France*

CAROL E. PARKER • *University of Victoria - Genome British Columbia Proteomics Centre, University of Victoria, Victoria, BC, Canada*

C. PASTOR-VARGAS • *Department of Immunology, Instituto de Investigacion Sanitaria Fundacion Jimenez Diaz (IIS-FJD), Madrid, Spain*

RAÚL PÉREZ • *Nanotechnology Platform, Barcelona Scientific Park, Barcelona, Spain*

ANDREW J. PERCY • *University of Victoria - Genome British Columbia Proteomics Centre, University of Victoria, Victoria, BC, Canada*

FLORENCE PINET • *INSERM, U744, Lille, France; Institut Pasteur de Lille, Lille, France; University Lille Nord de France, IFR142, Lille, France; Centre Hospitalier Régional et Universitaire de Lille, Lille, France*

MARIA POSADA-AYALA • *Department of Immunology, Instituto de Investigacion Sanitaria Fundacion Jimenez Diaz (IIS-FJD), Madrid, Spain*

PRISCILA RAMOS-MOZO • *Vascular Research Laboratory, IIS, Fundación Jiménez Díaz ands Autónoma University, Madrid, Spain*

JIAHUA SHI • *School of Chemical and Biomedical Engineering, College of Engineering, Nanyang Technological University, Singapore, Singapore*

KEE YANG TAN • *School of Chemical and Biomedical Engineering, College of Engineering, Nanyang Technological University, Singapore, Singapore*

JANE YI LIN TAN • *School of Chemical and Biomedical Engineering, College of Engineering, Nanyang Technological University, Singapore, Singapore*

XIAOLING TANG • *School of Chemical and Biomedical Engineering, College of Engineering, Nanyang Technological University, Singapore, Singapore*

JESUS VAZQUEZ • *Cardiovascular Proteomics Laboratory, Centro Nacional de Investigaciones Cardiovasculares (CNIC), Madrid, Spain*

FERNANDO VIVANCO • *Department of Immunology, IIS-Fundacion Jimenez Diaz, Madrid, Spain; Department of Biochemistry and Molecular Biology I Universidad Complutense, Madrid, Spain*

JIANHUA ZHANG • *School of Chemical and Biomedical Engineering, College of Engineering, Nanyang Technological University, Singapore, Singapore*

ALEJANDRO ZIMMAN • *Department of Molecular Cardiology, Lerner Research Institute, Cleveland Clinic, Cleveland, OH, USA*

IRENE ZUBIRI • *Department of Immunology, Instituto de Investigacion Sanitaria Fundacion Jimenez Diaz (IIS-FJD), Madrid, Spain*

Chapter 1

Vascular Proteomics

Maria G. Barderas, Fernando Vivanco, and Gloria Alvarez-Llamas

Abstract

Cardiovascular diseases constitute the largest of death in developed countries, being atherosclerosis the major contributor. Atherosclerosis is a process of chronic inflammation, characterized by the accumulation of lipids, cells, and fibrous elements in medium and large arteries. There is a continuum in atherosclerotic cardiovascular pathology that extends from the initial endothelial damage to diseases such as angina, myocardial infarction, and stroke. The extent of inflammation, proteolysis, calcification, and neovascularization influences the development of advanced lesions (atheroma plaques) on the arteries. Plaque rupture and the ensuing thrombosis cause the acute complications of atherosclerosis, i.e., myocardial infarction and cerebral ischemia. Thus, identification of early biomarkers of plaque unstability and susceptibility to rupture is of capital importance in preventing acute events. In recent years proteomics has been successfully applied to study proteins involved in these pathological processes. Thus, proteomic studies have been carried out focusing on different elements such as vascular tissues (arteries), artery layers, cells looking at proteomes and secretomes, plasma/serum, exosomes, lipoproteins, and metabolites. This chapter will provide an overview of latest advances in proteomic studies of atherosclerosis and related vascular diseases.

Key words Atherosclerosis, Vascular proteomics, Biomarkers, Serum, Plasma, Urine, Secretome, Metabolomics, MS imaging, Systems biology

1 Introduction

1.1 Atherosclerosis Development and Clinical Needs

Atherosclerosis, the most frequent etiology of cardiovascular disease (CVD), is a disease of blood vessels presenting a wide spectrum of pathological manifestations ranging from simple thickening and narrowing of the blood vessels to catastrophic coronary arterial occlusion and myocardial infarction. All atherosclerotic pathologies share the same cardiovascular risk factors and share pathogenic mechanisms. There is a continuum in atherosclerotic cardiovascular pathology that extends from the first endothelial damage to diseases such as angina, myocardial infarction, and stroke. Endothelial dysfunction induced by hypercholesterolemia is an initial step in atherosclerosis (1). The first endothelial dysfunction produces a decrease in nitric oxide production with a trend to

Fernando Vivanco (ed.), *Vascular Proteomics: Methods and Protocols*, Methods in Molecular Biology, vol. 1000, DOI 10.1007/978-1-62703-405-0_1,

vasoconstriction, thrombosis, and lipid accumulation in the arterial wall. All these factors lead to atherosclerotic plaque progression. Endothelial dysfunction is associated with high oxidative stress and oxidized low-density lipoproteins (LDL), which influences monocyte adhesion and migration into the subendothelial space, modification into macrophages, and the ensuing formation of foam cells. This results in the development of necrotic/lipidic cores within the intima of arteries at particular site in the circulation. These lesions form in the settings of a preexisting intimal hyperplasia characterized by the proliferation of VSMC within the intima. In advanced lesions, necrosis of macrophages and VSMC results in a lipid-rich core covered by a fibrous cap, which protects the lesions from rupture and consists mainly of collagen and extracellular matrix (ECM) proteins, synthesized by vascular cells. Plaque rupture, resulting from inflammatory activation and MMPs secretion, and the ensuing thrombosis commonly cause the most acute complications of atherosclerosis such as an acute coronary syndrome (ACS), i.e., myocardial infarction (in coronary arteries), cerebral ischemia (in the brain's irrigation system, i.e., carotid), or intermittent claudication and gangrene that jeopardize limb viability. In this context, multiple cell types (macrophages, vascular smooth muscle cells (SMC), and inflammatory cells, such as growth factors, connective tissue constituents, pro- and anticoagulants, lipid-associated proteins, metabolic regulators, and tissue enzymes) are involved, and the proteins and metabolites from these cells are likely to contribute to the pathogenesis of atherosclerosis.

In the clinical setting, the big issue is how to act on time. The plaque formation is a silent and asymptomatic process until progression reaches certain point when fatal events occur. In this context, there is urgent need to find out novel biomarkers of practical value for clinical intervention which, alone or combined with existing ones, allow cardiovascular risk prediction and early diagnosis at individual level. The omics platforms arise as disciplines providing a wide picture of proteins, peptides, and metabolites without preselection of potential targets (2).

2 Biomarkers of Disease

A biomarker is a characteristic that is "objectively measured and evaluated as an indicator of normal biological processes, pathogenic processes or pharmacological responses to therapeutic intervention" (3). The ideal biomarker must be accepted by the patient, easy to interpret, and able to explain a reasonable proportion of the outcome. Accuracy, reproducibility, availability, feasibility of implementation into the clinical settings, and specificity are additional characteristics to be fulfilled, and, in this sense, panels of biomarkers are gaining acceptance instead of individual molecules (4).

Candidate biomarkers should be carefully validated in a wide and different cohort of samples from those used in the discovery phase. The translation of the discovered biomarker into a routine clinical use is a step forward with no less difficulty, requiring the collaboration of the research laboratory, the diagnostics industry, and the clinical laboratory (5). Once gaining the battle of technology and being able to reach the sensitivity demanded by the perfect specific candidates to be detected, the last issue is finding suitable application, proving clinical relevance, and gaining industry acceptance. It will be then that the benefit for patients, industry, and society has reached its maximum expression.

The perfect proved candidate for biomarker of atherosclerosis and cardiovascular risk has not been discovered yet. Classical risk factors (hypertension, LDL-cholesterol, aging, smoking, male gender, among others) are not enough to act on time and prevent a fatal event. They have been put together in an algorithm widely used by clinicians to calculate 10-year risk of having cardiovascular adverse outcomes, the Framingham Risk Score (6), according to which stratification is built with three categories of higher probability of having cardiovascular events in the next 10 years: low (<10 %), intermediate (10–20 %), and high risk (>20 %). To date, several soluble molecules are used by clinicians to predict future cardiovascular events, including C-reactive protein (CRP), B-type natriuretic peptides, and cardiac troponins (cTnI, cTnT). CRP levels are a consequence of an acute-phase response, and special care has to be taken with possible underlying infectious or inflammatory diseases that may produce false-positive results. Cardiac troponins cTnI and cTnT are considered the most robust biomarkers in detection of myocardial injury, thus acute myocardial infarction (AMI) diagnosis, as well as in risk stratification of ACS (7). Elevated levels of both B-type natriuretic peptides have been correlated by several clinical trials with heart failure (HF) (8), making these peptides the prevalently selected biomarkers for HF diagnosis. Despite of the existence of these useful biomarkers of heart damage and heart failure, no early diagnosis biomarkers are available to date that may undoubtedly predict future events on healthy subjects. In this sense, the omics approach (proteomics and metabolomics) has emerged as a powerful tool, since it constitutes an undirected approach where the proteome or metabolome of a sample is analyzed as a whole.

3 A Step Forward: "Omics" Multi-target

In this moment it is very important to know what exactly "omics" mean. "Omics" refers to innovative technologies platforms formed for four main types of measurements that commonly performed: genomic, transcriptomic, proteomic, and metabolomic (Fig. 1).

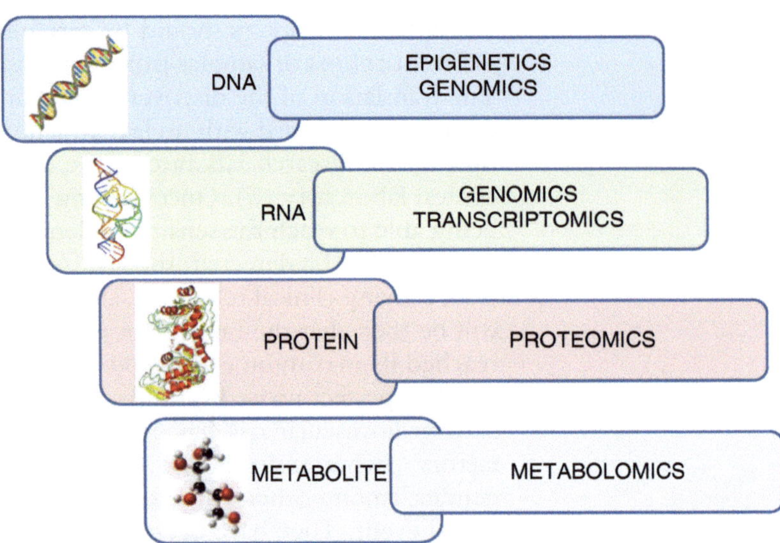

Fig. 1 The omics family

These recently developed techniques give us the ability to detect and identify many different molecules that are present and expressed in the body and, therefore, also in the cardiovascular system. Each of these four is distinct and offers a different perspective on the process underlying disease initiation and progression as well as on ways of predicting, preventing, or treating disease. Molecules such as DNA, RNA, proteins, peptides, lipids, and metabolites are detected and measured in different biological compartments such as whole blood, plasma, urine, and tissue. The "omics" technology allows the generation of copious amounts of data at multiple levels of biology from gene sequence and expression to protein and metabolite patterns underlying variability in cellular networks and function of whole organ system (9, 10). While the 90th was named as the "decade of brain," we are now in "the decade of measurements." This signals a new era in how we approach the scientific inquiries (11).

The "omics" technologies can be used to gain a "system-wide" understanding of many important biomedical processes. Furthermore, "omics" are becoming very useful and essential in order to (1) amplify the throughput in the process of research of new potential biomarkers, (2) increase the capacity to rapidly and inexpensively diagnose a disease, (3) monitor the effectiveness of treatment, and (4) identify the recurrence of disease far earlier than was once possible. In addition, these tools are opening up new approaches to drug development. On the other hand, with them, the experiments would be carried out to obtain data in order to test the study hypothesis, but, asking a question in an initial step of the research, it is not necessary or a prerequisite.

A step forward "in omics cardiovascular science" focuses to the whole set of molecules (DNA, RNA, proteins, peptides, lipids, and/or metabolites) without preselection of targets. The so generated data are not individual, referred to a unique molecule, but global, describing hundreds or thousands of results simultaneously. This fact is due to the ability and sensitivity of the new "omics" techniques to characterize tens, hundreds, or thousands of molecular species in each run and in a short period of time. As results, these produce the generation of profiles or data sets which reflect the situation of the analyzed sample. One of the main things to keep in mind is based in the different methodological approaches that are currently available, which means that we have to take into account different considerations: (a) the characteristics of the analytes to investigate (i.e., peptides, proteins, metabolites, lipids); (b) the performance offered by the technological platform in terms of sensitivity, selectivity, specificity, linear dynamic range, and throughput; and (c) the step in the biomarkers research pipeline to approach (discovery or validation). In the discovery phase, gel-based platforms (2D-DIGE) and liquid chromatography (nLC-MS/MS) setups are most commonly used for proteins analysis (12, 13), although the combination of capillary electrophoresis with mass spectrometry (CE-MS) for peptidome analysis is gaining popularity (14). Metabolome differential analysis is currently approached by LC-MS, gas chromatography on-line coupled to mass spectrometry (GC-MS), and nuclear magnetic resonance (NMR) (15, 16). For those pathologies where spatial distribution analysis of proteins, peptides, and metabolites can be useful, mass spectrometry imaging (MSI) is the platform of choice (17, 18). Once the potential biomarker candidate has been discovered, the next step of validation in a different cohort of samples should be approached. Apart from Western blot or Elisa, the analysis by selected reaction monitoring (SRM) is being increasingly established in current proteomics platforms. This strategy is typically performed in triple quadrupole configuration MS instrument, being able to monitor and quantify, simultaneously, hundreds of molecules per sample through the measurement of specific fragments coming from the proteins/metabolites of interest (19).

4 Sample Sources: The Advantages of Sub-proteomes

One of the most important things, in order to start a scientific research, is to know where we have to look for.

4.1 Biological Fluids

Plasma, serum, and urine are the most commonly used biological matrices in cardiovascular research, due to their availability and clinical relevance, as a source of potential biomarkers. Plasma is one of the best clinical sample in terms of diagnosis and prognosis, due to several advantages including low cost, noninvasiveness, and easy access (20). Furthermore, it presents a direct communication with almost all body cells, which release, at least, a part of their content into them (plasma/serum). For this reason, human plasma/serum is a rich source of molecules (proteins, metabolites, etc.) which reflect the physiological or clinical status of patients. However, the plasma proteome is very complex and presents a wide dynamic range of proteins (more than ten magnitude orders) that make its proteomic analysis very challenging, because high-abundance proteins tend to mask those of lower abundance (21). In other words, only 20 major proteins comprise 99 % of the plasma proteome and the rest of the proteins making up 1 % of the plasma content. Hence, it is essential to perform a pre-fractionated method; immunodepletion is the most common technique and is based on the action of specific antibodies (22). Among these, multiple affinity removal columns (MARC) is the most effective method, because it simultaneously removes multiple abundant proteins (23, 24). In the case of metabolites studies, a pretreatment with an organic solvent is required to remove all the proteins before starting the metabolomic analysis.

Urine is produced by renal filtration of the plasma, and it is widely considered as one of the most important samples for diagnosis, as it does not only contain many plasma components but also the catabolic products of different metabolic pathways. Moreover, the collection of urine is very simple, without any need for sophisticated or invasive procedures. Approximately 70 % of proteins in the normal human urinary proteome are kidney originated, whereas the remaining 30 % are derived from plasma proteins (25, 26).

Urine samples also meet the characteristics of an ideal biological matrix for discovering candidate metabolites. Unfortunately, there are several factors that play an important influence in the study of metabolomic: diet, physical exertion, stress, etc. These could produce urinary profiles with a big variability that must be carefully evaluated when using this sample for biomarker discovery.

Bronchoalveolar lavage, synovial fluid, saliva, and amniotic or cerebrospinal fluids are less common but can be easily related to a particular disorder and constitute more specific information sources. In this sense, these can be already considered as specific sub-proteomes/submetabolomes compared to whole plasma or urine. They account with the advantage of tremendously diminishing proteins abundance dynamic range, enormously facilitating the analysis from the technological side.

4.2 Tissue

Tissue offers a great promise to provide a clinical diagnosis and prognosis information that cannot be obtained from genomics or serum/plasma biomarkers. Today, there is an important need to develop standardized protocols and novel technologies that can be used in the routine clinical settings for seamless collection and immediate preservation of tissue in the search for biomarkers molecules (genes, proteins, peptides, lipids, metabolites). In this sense both clinicians and researchers are implicated. Normally, tissue is frozen in dry ice or liquid nitrogen or fixed to avoid preservation for later analyses, except when we want to study its secretome. A multitude of known and unknown variables can influence the stability of tissue molecules: temperature, pH, hypoxia, dehydration, RNAse activity, proteinases activity, ex vivo stress, etc. (27). For this reason the protocol to follow must be the strictest as possible: the recommended elapsed time is between 20 min and 2 h from surgery to stabilization. Finally the study of the tissue can be addressed according to three main samples sources: (1) whole tissue, (2) tissue subparts (i.e., layers, cells isolated, etc.), (3) tissue secretome.

4.2.1 Cells

Sub-fractionation of tissue in its structures or cellular components seems to be a complementary strategy to whole tissue analysis, in order to determine the specific contribution of whole tissue analysis and to determine specific contribution of these cells/structures in the pathogenesis of disease. The different structures from a histology section of a tissue can be isolated by means of microscopical dissection. Manual microdissection may be applied with the use of microscopical needle under an optical microscope that irradiates with a laser beam of a thermoplastic membrane in contact with the tissue, laser capture microdissection (LCM). Laser microdissection (LMD) (28) is a more accurate technique for the isolation of tissue regions, cells, or even subcellular fractions.

An alternative methodology to simplify tissue complexity and to study its cellular components is to enzymatically digest the tissue and separate cells for subsequent analysis. This can be done by cell sorting methodologies or by specific explant culturing. In the first one, specific cell populations can be separated in a flow cytometer and directly analyzed by proteomic techniques. The main problem in this workflow would be the high amount of tissue necessary to obtain enough cells for an analysis, which sometimes is impossible to extract, especially when dealing with biopsy material. Culturing sorted cells is an option but constitutes a more complicated approach than explants culturing, since sorted cells may be damaged in the cytometer, which involves lesser subculturing capacity. In contrast, non-sorted cell suspensions should be cultured under specific conditions to favor a certain cell type, which sometimes may imply contamination by other cells. The main drawback of subculturing methodologies is that the cells may lose its in vivo phenotype within the culture, since the environmental conditions are completely altered in the in vitro situation. An additional step in the sub-fractionation of the tissue would be to isolate cell structures/organelles. Conventional ultracentrifugation can be applied to separate membranes, nuclei, and cytosolic fractions for subsequent proteomic analysis.

4.2.2 Secretome

This term comprises the subset of proteins that are actively released by cells or tissue in the extracellular compartment as consequence of the normal metabolism or in response to some stimuli. As such, it is a powerful source of key molecules involved in pathogenesis development, individual response to pharmacological intervention, or recovery status. Secretome studies are increasing in the last years as it provides an accurate model of the in vivo situation and it represents a sub-proteome of serum/plasma, showing a much narrower proteins concentrations dynamic range which enormously facilitates detection of minor proteins whose identification is otherwise obscured by high-abundance plasma proteins. Furthermore, secretome-based research will favor the detection of novel proteins or, at least, known molecules with new implications in the disease under study. The most perfect approximation to the real situation is the study of the in vivo secretome. In this research line, one of the few proteomics studies was carried out by implanting capillary ultrafiltration probes into tumor masses induced in mice (29, 30). However, this is not always feasible and the use of tissue explants (ex vivo approach) represents a compromise solution which gives information about secretory molecules coming from all tissue components as result of cross-talk between them and approaches the physiological situation better than the cell culture. Two key points to consider when working directly with tissue in culture are (a) the need to ensure that all detected proteins are truly coming from the tissue and not "contaminants" derived from plasma and (b) validation of identified proteins as secreted. A metabolic labeling approach allows differentiation between proteins synthesized by the tissue (labeled) and contaminating proteins from blood which remain unlabelled (29). In this sense, an optimized culture protocol should be developed to maximize label incorporation into proteins including a series of medium changes during the initial hours of culture, followed by an extended step of tissue culture. In any case, label incorporation is influenced by the rate of synthesis of each particular protein, which may condition the number of labeled proteins that can be detected at a particular time point. Incorporation of the label by a protein validates tissue origin but does not necessarily imply "intentional" secretion. Once synthesized by the tissue, the release of a protein into the media could be attributed to damage-induced tissue leakage and results in detection of intracellular proteins. One can assume that it is very challenging to totally avoid the presence of intracellular proteins, as cell lysis always takes place during cell/tissue culture. However, optimum culture conditions may favor secreted proteins enrichment.

5 Proteomics in Cardiovascular Disease

Proteins are the essential effectors of genes, and they are important keys in the maintaining of cell, tissues, and organs. Consequently, proteins are widely used in different clinical tests for both diagnosis and prognosis of disease and can also contribute to study its evolution.

5.1 Proteins as Diagnostic Biomarkers

Despite the high prevalence of CVD in industrialized countries, there is currently not a rapid diagnostic assay for use in the emergency setting, which will allow a better management of the acute phase. Specifically, the formation of the atherome plaque is the causative factor that leads to coronary artery disease (CAD) (myocardial infarction and angina), peripheral vascular disease (PAD), and cerebral vascular disease (ischemic stroke). In all cases, the process is asymptomatic so that the discovery of potential biomarkers is an urgent need.

5.2 Classical Markers of Vascular Disease

The most extensively studied potential biomarker to date has been CRP. Although it could be of use in some risk situations, it appears to have only a moderate predictive value (31), and it has not gained wide acceptance in clinical practice. In addition to CRP, many different proteins, mainly involved in inflammation, have been studied in recent years as potential candidates for risk prediction. Among these are CD40L, monocyte chemoattractant protein-1 (MCP-1), adhesion molecules, myeloperoxidase, and several interleukins. Nevertheless, none of them has been consistently demonstrated to add predictive value to the clinical variables used in the clinical practice, and, in most cases, there are no commercially available standardized assays (32). One of the limitations of biomarker research is that, until now, each study may analyze data from thousands of patients but only focuses on a small number of proteins. The new proteomic approaches will probably represent a considerable advance by providing the investigators with the possibility of exploring hundreds of proteins at once and identifying new unforeseen biomarkers (33, 34).

5.3 Proteomic Approaches for the Study of Atherosclerosis

Until recently, the usual approach for the study of atherosclerosis has been to study the role of a candidate protein supposedly involved in the formation or progression of the atherosclerotic lesion. However, with the appearance of the new proteomic techniques of protein separation (2-DE, 2D-DIGE, multidimensional liquid chromatography) and their identification by mass spectrometry (MS), the evaluation of thousands of proteins at once is now possible. At present it is possible to perform a differential proteomic approach on a variety of biological samples, including cells, tissues, or biological fluids. In the context of biomarker discovery, biological fluids, such as plasma or urine, represent the most logical compartment for investigation because of their easy access.

5.4 Differential Expression of Proteins by Atherosclerotic Lesions

5.4.1 Biological Fluids

The initial 3,020-protein subset of the total proteins identified by the HUPO Plasma Proteome Project pilot phase (PPP) were researched in the literature for relevance to cardiovascular function and disease, resulting in at least 345 of them implicated, divided in eight different categories (markers of inflammation and CVD, vascular and coagulation, signaling, growth and differentiation, cytoskeletal, transcription factors, channel and receptor proteins, and heart failure and remodeling) (35). In this sense, Brea et al. (36) analyzed serum samples by 2-DE from ischemic stroke patients previously depleted in the 12 more abundant plasma proteins. Results showed altered haptoglobin and serum amyloid A (SAA) expression depending on the operating atherothrombotic mechanism; higher serum levels of these proteins can predict atherothrombotic versus cardioembolic stroke. By SELDI-TOF-MS, β2-microglobulin was found to be elevated in plasma of patients with peripheral arterial disease (PAD) and correlates with severity. As an alternative to SELDI-TOF which enrich for subsets of proteins on the sample target surface, Ganesh et al. (37) have used a direct MS method to analyze serum in a cohort of patients with venous thromboembolism (VTE). Current noninvasive testing for VTE includes blood assays of D-dimer. Although this assay provides a high negative predictive value it suffers for diminished specificity. Thus, authors developed a direct MS and computational approach to determine whether protein expression profiles would predict diagnosis. The data were confirmed by gel electrophoresis of serum proteins. Proteins identified include actin, α-1-B-glycoprotein, CD5 antigen-like, complement 4A and 9 proteins, haptoglobin, hemopexin, IgA heavy chain, leucine-rich α2 glycoprotein 1, myosin heavy chain, platelet coagulation factor XI, plasma kallikrein B1 precursor, and proapolipoprotein. Since these proteomic markers appear to add specificity to currently available blood assays, they could be used in combination with other methods, including radiographic and ultrasound evaluation, in order to obtain optimal sensitivity and specificity for the diagnosis of VTE. Recently, Liu et al. (38) have used a label-free proteomic method with LC-MS/MS (shotgun) to investigate the differences in the protein profiles between the nondiabetic ($n=5$) and diabetic ($n=5$) sera. They analyzed complete serum (without depletion of abundant proteins) in order to avoid protein losses and generation of potential artifacts. As a result, 68 proteins were found significantly overrepresented in the diabetic serum, from which 12 belonged to the complement system.

Some researchers prefer urine instead of plasma for proteomic analysis since they found urine to be stable against proteolytic degradation, contains a low concentration of irrelevant proteins and can be collected noninvasively. For example, Zimmerli et al. (39) examined a total of 359 urine samples, from 88 patients with severe CAD, and 282 controls by capillary electrophoresis and ESI-TOF/MS.

Although CE is not yet a routinary tool for massive separation of proteins, its utilization is progressively increasing (40), and in this work, more than 1,000 polypeptides per sample were analyzed. A set of 15 peptides that define a characteristic CAD signature panel was identified. Curiously the majority of the identified polypeptide patterns able to discriminate between the presence and the absence of disease consist in fragments of collagen α-1 (I) chains and collagen α-1 (III). These collagens, types I and III, are predominant proteins in the arterial walls and appear also together in the thickened intima of atherosclerotic lesions. A similar study has been subsequently reported evaluating the urine proteome pattern for its potential to reflect coronary artery atherosclerosis in symptomatic patients that confirm the previous results (41). Thus, the comparison in the polypeptide pattern obtained in these two separate studies supports that CAD can be reflected in specific polypeptide patterns in urine.

5.4.2 Tissue

A major obstacle for applying proteomic analysis to vascular pathology is the heterogeneous cellular composition of atherosclerotic plaques (endothelial cells, EC; vascular smooth muscle cells, VSMC; leukocytes; and erythrocytes). Thus, a proteomic analysis of the whole tissue (carotid endarterectomy) probably identifies mainly constitutive proteins and may miss underrepresented proteins which could serve as biomarkers. Hence, other strategies have been adopted which are described below.

Martinet et al. (42, 43) have screened tissue lysates from human carotid atherosclerotic samples and healthy mammary arteries by Western array using 823 monoclonal antibodies. They reported seven proteins with >5-fold relative expression difference, one of which was the apoptosis-linked gene 2 (ALG-2), a positive mediator of apoptotic cell death. Since apoptosis could be implicated with atherosclerotic plaque instability, the decreased levels observed for ALG-2 suggest a new survival mechanism in human atherosclerotic plaques.

When 2-DE profiles of whole-mounted advanced human stable lesions were compared with plaques containing a thrombus (44), 71 spots were present exclusively in stable plaques and 29 in thrombus-containing plaques. This analysis revealed the expression of six isoforms of α_1-antitrypsin (ATT) in advanced plaques, one of which was uniquely expressed in thrombus-containing plaques.

Using antibody microarrays, Slevin et al. (45) have identified novel proteins associated with the development of unstable human carotid plaques. They compared protein expression in carotid endarterectomy samples histologically defined as stable and unstable. In endothelial cells and VSMC, several proteins were overexpressed including caspase-9, TRAF4, and topoisomerase II-α. In addition, cell-signaling proteins c-src, G-protein-coupled receptor kinase interacting protein (GIT1), and c-Jun N-terminal kinase (JNK) were upregulated in endothelial cells from the same areas.

5.4.3 Secretome

Most studies are focused on the secretome obtained from in vitro cell cultures, assuming that such cells' behavior well simulates the in vivo condition (46). The most perfect approximation to the real situation is the study of the in vivo secretome. We have reported two approaches for the study of atherosclerotic plaque secretome in the search of potential biomarkers of atherosclerosis (47–50). The first of all consisted in comparing the secretome from normal and pathological arteries using a differential proteomic approach to identify new biological markers potentially released by the arterial wall into the plasma. Incubation of samples versus control endarteries in a protein-free culture medium allowed us to harvest separately the proteins released from pathological and healthy areas. Using this approach, we could see that in comparison with healthy mammary arteries, atherosclerotic plaques release less phosphorylated HSP27 (51), which was further confirmed by Western blot and ELISA. The biological significance of decreased HSP27 plasma levels in human atherosclerosis has been described too (52). On the other hand, secretomes of human radial artery segments (control) were compared with carotid sections with atherosclerotic plaque of variable severity (noncomplicated and ruptured plaques with thrombus) in a 2-DE approach (53). Three biological replicates were analyzed per group, and a total of 64 proteins were identified in the three replicates of at least one group. Among them, 14 secreted proteins have not been previously reported in plasma. By means of label-free MS/MS-based quantification, the comparison of atheroma plaque coronary, preatherosclerotic coronary, and mammary secretomes was performed. Four proteins were commonly found to be highly released from mammary versus coronary arterial tissue: gelsolin, vinculin, lamin A/C, and phosphoglucomutase 5.

6 Metabolomics in Cardiovascular Disease

Metabolites are small molecules (>1 kDa) that participate in general metabolic reactions and that are required for the maintenance, growth, and normal function of a cell. Metabolomics refer to the systematic analysis of metabolites in a biological system, as well as the monitoring of changes in the metabolome of a biofluid, cell culture, or tissue sample following perturbation (54, 55). From a clinical perspective, the study of metabolic changes that occur in response to different physiological processes will help establishing the mechanisms underlying the disease. By performing global metabolite profiling, also known as untargeted metabolomics, new discoveries linking cellular pathways to biological mechanism are being revealed. There are several analytical strategies that can be used to analyze the metabolome (56), such as NMR Fourier transformation infrared spectroscopy (FT-IR) (57) and mass spectrometry (MS) coupled to separation techniques such as high-performance liquid chromatography (HPLC), gas chromatography (GC), or capillary electrophoresis (CE). High-field ^1H NMR (58) is a robust, nondestructive technique that does not require prior separation of the analytes or sample treatment, and it provides detailed information on molecular structure. However, one of the main limitations of NMR is the poor sensitivity. GC-MS provides an extraordinary resolution, able to separate structurally similar compounds (i.e., fatty acids, organic acids, steroids, diglycerides, sugars, and sugar alcohols). However, this technique requires the analyte to be volatile and thermally stable, and it may require chemical derivatization prior to the chromatographic separation. For those metabolites that are not volatile and which cannot be derivatized, LC is the separation technique of choice. LC-MS can analyze a much wider range of chemical species (polar and nonpolar metabolites) with ample selectivity and sensitivity. Apart from reversed-phase chromatography (RP-LC), which is widely used in metabolomics applications, hydrophilic interaction chromatography (HILIC) is a complementary approach suitable for very polar metabolites (nonvolatile). NMR and MS can be used for both main strategies of metabolic studies: profiling (nontargeted) and fingerprinting (targeted). The profiling approach focuses on the analysis of a group of metabolites related to a specific metabolic pathway selected beforehand (59–61). Metabolic fingerprinting does not aim to identify the entire set of metabolites but rather to compare patterns or fingerprints of metabolites that change in response to a disease state, pharmacological therapies, or environmental alterations. Compilations of metabolomic studies in CVD have been recently published (15, 16). Metabolic changes associated to atherosclerosis were investigated through NMR and GC-MS metabolite profiling (62–64). Plasma and urine samples from atherosclerotic and control rats have been compared by ultrafast liquid chromatography coupled to ion trap/time-of-flight (IT-TOF) mass spectrometry, identifying 12 metabolites as potential biomarkers in rat plasma and 8 metabolites in rat urine (65). The myocardial metabolic response has been investigated in CAD and left ventricular dysfunction (LVD) patients, both at baseline and following ischemia-reperfusion (I/R) (66), finding some citric acid metabolites depressed in acute ischemia and myocardial disease (67). Targeted metabolic profiling was also applied to investigate blood metabolite alterations produced by planned myocardial infarction (PMI) (68). The metabolomic fingerprint of non-ST-segment elevation acute coronary syndrome (NSTEACS) patients, stable atherosclerosis patients, and healthy patients was comparatively analyzed by GC-MS plasma analysis (69), and the polypeptide fraction from urine and plasma, analyzed by CE-MS, was used to discriminate between CAD and non-CAD patients with clinical symptoms (70). The increasing popularity and acceptance of metabolomics research will undoubtedly result in a near future gain of knowledge of CVD metabolism.

7 Emerging Mass Spectrometry Imaging: High Potential in CVD

MSI is an emerging platform developing quickly, as it is able to provide with a real "picture" of local distribution of proteins and metabolites directly in the tissue. It can be considered as a multiplex of conventional immunohistochemistry (IHC) in the sense that it has the capability to measure simultaneously non-targeted proteins, peptides, and metabolites present in a tissue section without missing spatial information (17). An additional advantage relies on the fact that it is not necessary to know beforehand the target protein to select the appropriated antibody for staining as in IHC. By this technique, thousands of molecules are being measured at the same time without preselection, and it is possible to detect isoforms of a certain protein with potential different roles in disease. Spatial resolution is conditioned by the ionization technique, sensitivity, mass resolution, and the capability to measure different classes of biomolecules at the same time by the mass spectrometer characteristics, which must be able to simultaneously detect a wide broad range of ions (from low to high Mw). There are three key steps in MSI: sample preparation, desorption/ionization, mass analysis, and image acquisition. Sample preparation is a crucial step and special care has to be taken to avoid molecular diffusion. One of the main limitations is the ion suppression effect, particularly present in such a complex sample as it is human tissue. The presence of lipids, carbohydrates, and salts may promote adduct formation, affecting co-crystallization of biomolecules with matrix and influencing the quality of the MS spectra and the number of detected molecules. In the particular case of vascular tissue, hemoglobin from blood may have a deleterious effect on the rest of the signals as hemoglobin chains are easily ionizable and may compromise other proteins signals (71). In this context, few studies have been reported so far in the cardiovascular field. Atherosclerotic lesions from aortic roots of apoE-deficient mice and from human femoral arteries with PAD were profiled by MSI, and differential molecules related to specific plaque areas were identified (72). MALDI imaging was also applied to chick heart tissue sections acquired from fixed and paraffin-embedded samples (73), and an in vivo rat model of myocardial infarction was investigated by MSI, finding phospholipid markers in areas of infarction (74). Particular aspects concerning methodology, fields of application, and key issue to take into account related to sample preparation and tissue preservation have been recently reviewed (75, 76). Molecular spatial distribution maps are so generated opening a new field of research.

It has been applied to study changes in protein expression levels, molecular distributions associated with a range of pathologies and pharmaceuticals, and their metabolites distribution within organs and complete animals. An excellent special issue of *Journal of Proteomics* ((75(16), 2012) dedicated to MS imaging has been very recently published.

8 Systems Biology

In the whole context, atherosclerosis is a complex multifactorial disease. Classical studies of individual components (genes, proteins, metabolites) are necessary to establish the function of such components, but they are not sufficient to explain complex processes. A more global analysis is necessary in which the activities of all relevant molecules are integrated and thus could provide a complete view of how they function together. These global approaches, named systems biology, enable the identification of networks of proteins or metabolites associated with CVD in general and with atherosclerosis in particular (77, 78). The ultimate clinical presentation of CVD results from the interaction of multiple cell types (macrophages, endothelial cells, VSMC, lymphocytes) and organ systems (vascular, endocrine, adipose, liver, kidney, gastrointestinal) in which a myriad of interconnected proteins are expressed. Thus, systems-based approaches are very well suited for elucidating the high-order interactions underlying the process of atherosclerosis and providing a framework for the identification of potential biomarkers (79), novel drugs, and personalized treatments (80). Proteins are the ultimate expression of genes, and metabolites represent the end products of the genome and proteome, providing an instantaneous snapshot of the physiology of a cell, tissue, or organism. The omics platforms represent a range of opportunities to study biological systems as a whole and from different perspectives (Fig. 2). By means of the correct choice of technological and sampling approach, scientific research can be biased to a deeper comprehension of disease mechanisms themselves or key targets involved in pathology development. In all cases, different techniques employed and molecular types (proteins or metabolites) chosen to be investigated provide with complementary information. Bioinformatics tools are able to generate biological networks, and by means of, i.e., ingenuity pathways analysis (IPA) (Ingenuity® Systems, www.ingenuity.com), top biofunctions can be obtained together with the associated significance (81). In a similar manner, STRING is a database of known and predicted protein interactions including direct (physical) and indirect (functional) associations (82).

SYSTEMS OF BIOLOGY

Biological samples in proteomics

Tissue and cell culture

Plasma/serum

Extravascular fluids

Urine

Biological samples in metabolomics

Tissue and cell culture

Plasma/serum

Extravascular fluids

Urine

DISCOVERY → QUANTIFICATION → BIOINFORMATIC ANALYSIS

Fig. 2 Biological samples used in proteomics and metabolomics

Acknowledgments

Work in the authors' laboratories has been supported by grants FIS PI11/01401.

References

1. Davignon J, Ganz P (2004) Role of endothelial dysfunction in atherosclerosis. Circulation 109:III27–III32
2. Alvarez-Llamas G, de la Cuesta F, Barderas MG, Darde V, Padial LR, Vivanco F (2008) Recent advances in atherosclerosis-based proteomics: new biomarkers and a future perspective. Expert Rev Proteomics 5:679–691
3. Biomarkers Definitions Working Group (2001) Biomarkers and surrogate endpoints: preferred definitions and conceptual framework. Clin Pharmacol Ther 69:89–95
4. Finley Austin MJ, Babiss L (2006) Commentary: where and how could biomarkers be used in 2016. AAPS J 8:E185–E189
5. Sturgeon C, Hill R, Hortin GL, Thompson D (2010) Taking a new biomarker into routine use—a perspective from the routine clinical biochemistry laboratory. Proteomics Clin Appl 4:892–903
6. Wilson PW, D'Agostino RB, Levy D, Belanger AM, Silbershatz H, Kannel WB (1998) Prediction of coronary heart disease using risk factor categories. Circulation 97:1837–1847
7. Apple FS, Collinson PO (2012) IFCC task force on clinical applications of cardiac biomarkers. Analytical characteristics of high-sensitivity cardiac troponin assays. Clin Chem 58:54–61
8. Di Angelantonio E, Chowdhury R, Sarwar N, Ray KK, Gobin R, Saleheen D et al (2009) B-type natriuretic peptides and cardiovascular risk: systematic review and meta-analysis of 40 prospective studies. Circulation 120:2177–2187
9. Nicholson JK, Lindon LC (2008) Systems biology: metabolomics. Nature 455:1054–1056

10. Wike RA, Mereedu RK, Moore JH (2008) The pathway less traveled: moving from candidate genes to candidate pathways in the analysis of genome-wide data from large scala pharmacogenetic association studies. Curr Pharmacogenomics Person Med 6:150–159
11. Ozdemir V, Suarez-Kurtz G, Stenne E, Somoggyi A, Kayaalp O, Kolker E (2008) Risk assessment and communication tools for genotype associations with multifactorial phenotypes: The concept of "edge effect" and cultivating an ethical bridge between omics innovations and society. J Integr Biol 13:43–62
12. Beer LA, Tang H, Barnhart KT, Speicher DW (2011) Plasma biomarker discovery using 3D protein profiling coupled with label-free quantitation methods. Mol Biol 728:3–27
13. Thakur SS, Geiger T, Chatterjee B, Bandilla P, Fröhlich F, Cox J, Mann M (2011) Deep and highly sensitive proteome coverage by LC-MS/MS without prefractionation. Mol Cell Proteomics 10:M110.003699
14. Mischak H, Coon JJ, Novak J, Weissinger EM, Schanstra J, Dominiczak AF (2009) Capillary electrophoresis—mass spectrometry as a powerful tool in biomarker discovery and clinical diagnosis: an update of recent developments. Mass Spectrom Rev 28:703–724
15. Barderas MG, Laborde CM, Posada M, de la Cuesta F, Zubiri I, Vivanco F, Alvarez-Llamas G (2011) Metabolomic profiling for identification of novel potential biomarkers in cardiovascular diseases. J Biomed Biotechnol doi: 10.1155/2011/790132.
16. Rhee EP, Gerszten RE (2012) Metabolomics and cardiovascular biomarker discovery. Clin Chem 58:139–147
17. McDonnell LA, Heeren RMA (2007) Imaging mass spectrometry. Mass Spectrom Rev 26: 606–643
18. Wang J, Balu N, Canton G, Yuan C (2010) Imaging biomarkers of cardiovascular disease. J Magn Reson Imaging 32:502–515
19. Lange V, Picotti P, Domon B, Aebersold R (2008) Selected reaction monitoring for quantitative proteomics: a tutorial. Mol Syst Biol 4:222
20. Veenstra TD, Conrads TP, Hood BL, Avellino AM, Ellenbogen RG (2005) Biomarkers: mining the biofluid proteome. Mol Cell Proteomics 4:409–418
21. Anderson NL (2005) Candidate-based proteomics in the search for biomarkers of cardiovascular disease. J Physiol 563:23–60
22. Wang YY, Cheng P, Chan DW (2003) A simple affinity spin tube filter method for removing high-abundant common proteins or enriching low-abundant biomarkers for serum proteomic analysis. Proteomics 3:243–248
23. Bjorhall K, Miliotis T, Davidsson P (2005) Interest of major serum protein removal for surface-enhanced laser desorption/ionization—time of flight (SELDI-TOF) proteomic blood profiling. Proteomics 5:307–317
24. Seam N, Gonzales DA, Kern SJ, Hortin GL, Hoehn GT, Suffredini AF (2007) Quality control of serum albumin depletion for proteomic analysis. Clin Chem 53:1915–1920
25. Thongboonkerd V, McLeish KR, Arthur JM, Kelin JB (2002) Proteomic analysis of normal human urinary proteins isolated by acetone precipitation or ultracentrifugation. Kidney Int 62:1461–1469
26. Thongboonkerd V, Malasit P (2007) Renal and urinary proteomics: current applications and challenges. Proteomics 5:1033–1042
27. Espina V et al (2008) A portrait of tissue phosphoprotein stability in the clinical tissue procurement process. Mol Cell Proteomics 7:1998–2018
28. Emmert-Buck MR et al (1996) Laser capture microdissection. Science 274:998–1001
29. Alvarez-Llamas G et al (2007) Characterization of the human visceral adipose tissue secretome. Mol Cell Proteomics 6:589–600
30. Hocking SL et al (2010) Intrinsic depot-specific differences in the secretome of adipose tissue, preadipocytes, and adipose tissue-derived microvascular endothelial cells. Diabetes 59:3008–3016
31. Danesh J, Wheeler JG, Hirschfield GM, Eda S, Eiriksdottir G, Rumley A, Lowe GD, Pepys MB, Gudnason V (2004) C-reactive protein and other circulating markers of inflammation in the prediction of coronary heart disease. N Engl J Med 350:1387–1397
32. Sidney C, Smith J, Anderson JL, Cannon RO, Fadl YY, Koenig W, Libby P, Lipshultz SE, Mensah GA, Ridker PM, Rosenson R (2004) CDC/AHA Workshop on markers of inflammation and cardiovascular disease application to clinical and public health practice: report from the Clinical Practice Discussion Group. Circulation 110:e550–e555
33. Mayr M, Mayr U, Chung YL, Yin X, Griffiths JR, Xu Q (2004) Vascular proteomics: linking proteomic and metabolomic changes. Proteomics 4:3751–3761
34. Marian AJ, Nambi V (2004) Biomarkers of cardiac disease. Expert Rev Mol Diagn 2004(4):805–820
35. Berhane BT, Zong C, Liem DA, Huang A, Le S, Edmondson RD, Jones RC, Qiao X, Whitelegge JP, Ping P, Vondriska TM (2005) Cardiovascular-related proteins identified in human plasma by the HUPO Plasma Proteome Project pilot phase. Proteomics 5:3520–3530

36. Brea D, Sobrino T, Blanco M, Fraga M, Agulla J, Rodriguez-Yañez M, Rodriguez-Gonzalez R, Perez de la Ossa N, Leira R, Forteza J, Davalos A, Castillo J (2009) Usefulness of haptoglobin and serum amyloid A proteins as biomarkers for atherothrombotic ischemic stroke diagnoses confirmation. Atherosclerosis 205(2):561–567
37. Ganesh SK, Sharma Y, Dayhoff J, Fales HM, Van Eyk J, Kickler TS, Billings EM, Nabel EG (2007) Detection of venous thromboembolism by proteomic serum biomarkers. PLoS One 2(6):e544
38. Liu RX, Chen HB, Tu K, Zhao SH, Li SJ, Dai J, Li QR, Nie S, Li YX, Jia WP, Wu JR (2008) Localized-statistical quantification of human serum proteome associated with type 2 diabetes. PLoS One 3:e3224
39. Zimmerli LU, Schiffer E, Zürbig P, Good DM, Kellmann M, Mouls L, Pitt AR, Coon JJ, Schmieder RE, Peter KH, Mischak H, Kolch W, Delles C, Dominiczak AF (2008) Urinary proteomic biomarkers in coronary artery disease. Mol Cell Proteomics 7:290–298
40. Coon J, Zurbig P, Dakna M, Dominiczak A, Decramer S et al (2008) CE-MS analysis of the human urinary proteome for biomarker discovery and disease diagnostics. Proteomics Clin Appl 2:964–973
41. Von zur Muhlen C, Schiffer E, Zuerbig P, Kellmann M, Brasse M, Meert N, Vanholder RC, Dominiczak AF, Chen YC, Mischak H, Bode C, Peter K (2009) Evaluation of urine proteome pattern analysis for its potential to reflect coronary artery atherosclerosis in symptomatic patients. J Proteome Res 8: 335–345
42. Martinet W, Schrijvers DM, De Meyer GR, Herman AG, Kockx MM (2003) Western array analysis of human atherosclerotic plaques. Down regulation of apoptosis-linked gene 2. Cardiovasc Res 60:259–267
43. Martinet W (2006) Western array analysis of human atherosclerotic plaques. Methods Mol Biol 357:165–178
44. Donners M, Verluyten MJ, Bouwman FG, Mariman E, Devrese B et al (2005) Proteomic analysis of differential protein expression in human atherosclerotic plaque progression. J Pathol 206:39–45
45. Slevin M, Elasbali AB, Turu M, Krupinski J, Badimon L et al (2006) Identification of differential protein expression associated with development of unstable human carotid plaques. Am J Pathol 168:1004–1021
46. Roelofsen H, Dijkstra M, Weening D, de Vries MP, Hoek A, Vonk RJ (2009) Comparison of isotope-labeled amino acid incorporation rates (CILAIR) provides a quantitative method to study tissue secretomes. Mol Cell Proteomics 8:316–324
47. Durán MC, Mas S, Martin-Ventura JL, Meilhac O, Michel JB et al (2003) Proteomic analysis of human vessels: application to atherosclerotic plaques. Proteomics 3:973–978
48. Durán MC, Martin-Ventura JL, Mohammed S, Barderas MG, Mas S et al (2007) Atorvastatin modulates the profile of proteins released by human atherosclerotic plaques. Eur J Pharmacol 562(1–2):119–129
49. Duran MC, Martín-Ventura JL, Mas S, Barderas MG, Darde V et al (2006) Characterization of the human atherome plaque secretome by proteomic analysis. Methods Mol Biol 357:141–150
50. Vivanco F, Martin-Ventura JL, Duran MC, Barderas MG, Blanco-Colio L et al (2005) Quest for novel cardiovascular biomarkers by proteomic analysis. J Proteome Res 4:1181–1191
51. Martín-Ventura JL, Duran MC, Blanco-Colio L, Meilhac O, Leclercq A et al (2004) Identification by a differential proteomic approach of heat shock protein 27 as a potential marker of atherosclerosis. Circulation 110:2216–2219
52. Martin-Ventura JL, Nicolas V, Houard X, Blanco-Colio L, Leclercq A et al (2006) Biological significance of decreased HSP27 in human atherosclerosis. Arterioscler Thromb Vasc Biol 26:1337–1343
53. de la Cuesta F, Barderas MG, Calvo E, Zubiri I, Maroto AS, Darde VM, Martin-Rojas T, Gil-Dones F, Posada-Ayala M, Tejerina T, Lopez JA, Vivanco F, Alvarez-Llamas G (2011) Secretome analysis of atherosclerotic and non-atherosclerotic arteries reveals dynamic extracellular remodeling during pathogenesis. J Proteomics 75(10): 2960–2971
54. Fiehn O, Kopka J, Dormann P, Altmann T, Trethwey R, Wilmitzer J (2000) Metabolite profiling for plant functional genomics. Nat Biotechnol 18:1157–1161
55. Nicholson JK, Lindon J, Holmes E (1999) Metabolomics: understanding the metabolic responses of living systems to pathophysiological stimuli via multivariate statistical analysis of biological NRM spectroscopic data. Xenobiotica 29:1181–1189
56. Dunn W, Bailey N, Johnso H (2005) Measuring the metabolome: current analytical technologies. Analyst 130:606–625
57. Harrigan G, LaPlante R, Cosma G, Cockerell G, Goodacre R, Maddox J et al (2004) Application of high-throughput Fourier-transform infrared spectroscopy in toxicology

studies: contribution to a study on the development of an animal model for idiosyncratic toxicity. Toxicol Lett 146:197–205

58. Lenz EM, Bright J, Wilson ID, Morgan SR, Nash AF (2003) A 1H NMR-based metabonomic study of urine and plasma samples obtained from healthy human subjects. J Pharm Biomed Anal 33:1103–1115
59. Musiek ES, Yin H, Milne GL, Morrow JD (2005) Recent advances in the biochemistry and clinical relevance of the isoprostane pathway. Lipids 40:987–994
60. Cho HJ, Kim JD, Lee WY, Chung BC, Choi MH (2009) Quantitative metabolic profiling of 21 endogenous corticosteroids in urine by liquid chromatography-triple quadrupole-mass spectrometry. Anal Chim Acta 632:101–108
61. Wang Z, Tang WH, Cho L, Brennan DM, Hazen SL (2009) Targeted metabolomic evaluation of arginine methylation and cardiovascular risks: potential mechanisms beyond nitric oxide synthase inhibition. Arterioscler Thromb Vasc Biol 29:1383–1391
62. Teul J, Ruperez FJ, Garcia A, Vaysse J, Balayssac S, Gilard V et al (2009) Improving metabolite knowledge in stable atherosclerosis patients by association and correlation of GC-MS and 1H NMR fingerprints. J Proteome Res 8: 5580–5589
63. Mayr M, Yusuf S, Weir G, Chung YL, Mayr U, Yin X et al (2008) Combined metabolomic and proteomic analysis of human atrial fibrillation. J Am Coll Cardiol 51:585–594
64. Chen X, Liu L, Palacios G, Gao J, Zhang N, Li G, Lu J, Song T, Zhang Y (2010) Plasma metabolomics reveals biomarkers of the atherosclerosis. J Sep Sci 33:2776–2783
65. Zhang F, Jia Z, Gao P, Kong H, Li X, Chen J et al (2009) Metabonomics study of atherosclerosis rats by ultra fast liquid chromatography coupled with ion trap-time of flight mass spectrometry. Talanta 79:836–844
66. Turer AT, Stevens RD, Bain JR, Muehlbauer MJ, van der WJ, Mathew JP et al (2009) Metabolomic profiling reveals distinct patterns of myocardial substrate use in humans with coronary artery disease or left ventricular dysfunction during surgical ischemia/reperfusion. Circulation 119:1736–1746
67. Zhao G, Jeoung NH, Burgess SC, Rosaaen-Stowe KA, Inagaki T, Latif S et al (2008) Overexpression of pyruvate dehydrogenase kinase 4 in heart perturbs metabolism and exacerbates calcineurin-induced cardiomyopathy. Am J Physiol Heart Circ Physiol 294:H936–H943
68. Lewis GD, Wei R, Liu E, Yang E, Shi X, Martinovic M et al (2008) Metabolite profiling of blood from individuals undergoing planned myocardial infarction reveals early markers of myocardial injury. J Clin Invest 118:3503–3512
69. Vallejo M, Garcia A, Tunon J, Garcia-Martinez D, Angulo S, Martin-Ventura JL et al (2009) Plasma fingerprinting with GC-MS in acute coronary syndrome. Anal Bioanal Chem 394:1517–1524
70. von Zur Muhlen C, Schiffer E, Zuerbig P, Kellmann M, Brasse M, Meert N et al (2009) Evaluation of urine proteome pattern analysis for its potential to reflect coronary artery atherosclerosis in symptomatic patients. J Proteome Res 8:335–345
71. Schwartz SA, Reyzer ML, Capriol RM (2003) Direct tissue analysis using matrix-assisted laser desorption/ionization mass spectrometry: practical aspects of sample preparation. J Mass Spectrom 38:699–708
72. Zaima N, Sasaki T, Tanaka H, Cheng XW, Onoue K, Hayasaka T, Goto-Inoue N, Enomoto H, Unno N, Kuzuya M, Setou M (2011) Imaging mass spectrometry-based histopathologic examination of atherosclerotic lesions. Atherosclerosis 217:427–432
73. Grey AC, Gelasco AK, Section J, Moreno-Rodriguez RA, Krug EL, Schey KL (2010) Molecular morphology of the chick heart visualized by MALDI imaging mass spectrometry. Anat Rec (Hoboken) 293:821–828
74. Menger RF, Stutts WL, Anbukumar DS, Bowden JA, Ford DA, Yost RA (2012) MALDI mass spectrometric imaging of cardiac tissue following myocardial infarction in a rat coronary artery ligation model. Anal Chem 84:1117–1125
75. Chughtai K, Heeren RM (2010) Mass spectrometric imaging for biomedical tissue analysis. Chem Rev 110:3237–3277
76. Seeley EH, Caprioli RM (2011) MALDI imaging mass spectrometry of human tissue: method challenges and clinical perspectives. Trends Biotechnol 29:136–143
77. Wheelock CE, Wheelock AM, Kawashima S, Diez D, Kanehisa M, van Erk M, Kleemann R, Haeggström JZ, Goto S (2009) Systems biology approaches and pathway tools for investigating cardiovascular disease. Mol BioSyst 5:588–602
78. Ramsey SA, Gold ES, Aderem A (2010) A systems biology approach to understanding atherosclerosis. EMBO Mol Med 2:79–89
79. Jain KK (2010) Technologies for discovery of biomarkers. In: The handbook of biomarkers. Springer, New York. Jain Pharma-Biotech, Basel, Switzerland

80. Lusis AJ, Weiss JN (2010) Cardiovascular networks: systems-based approaches to cardiovascular disease. Circulation 121:157–170
81. de la Cuesta F, Barderas MG, Calvo E, Zubiri I, Maroto AS, Darde VM, Martin-Rojas T, Gil-Dones F, Posada-Ayala M, Tejerina T, Lopez JA, Vivanco F, Alvarez-Llamas G (2012) Secretome analysis of atherosclerotic and non-atherosclerotic arteries reveals dynamic extracellular remodeling during pathogenesis. J Proteomics 75:2960–2971
82. Szklarczyk D, Franceschini A, Kuhn M, Simonovic M, Roth A, Minguez P, Doerks T, Stark M, Muller J, Bork P, Jense LJ, von Mering C (2011) The STRING database in 2011: functional interaction networks of proteins globally integrated and scored. Nucleic Acids Res 39:D561–8

Chapter 2

Laser Microdissection and Saturation Labeling DIGE Method for the Analysis of Human Arteries

Fernando de la Cuesta, Gloria Alvarez-Llamas, Aroa S. Maroto, Maria G. Barderas, and Fernando Vivanco

Abstract

Laser microdissection (LMD) is a novel methodology for noncontact isolation of tissue regions or cells for subsequent molecular analysis. Although it is an upcoming field, its combination with proteomics for differential analysis remains not very well explored, since amount of protein obtained after LMD is scarce. We have combined LMD arterial layer isolation with saturation labeling DIGE, successfully achieving differential analysis of healthy and pathological intima and media layers. Identification of differential spots could be performed in whole tissue extract as reference proteome, since studied regions are subproteomes of the aforementioned.

Key words Laser microdissection, Coronary artery, Radial artery, Aorta artery, Atherosclerosis, Proteomics, Saturation labeling DIGE

1 Introduction

Events provoked by coronary atherosclerosis constitute the main cause of death in developed countries. Although our knowledge of atherosclerosis pathogenesis is constantly growing (1), the underlying molecular mechanisms remain uncertain. Tissue analysis of coronary artery, where the pathology arises, provides interesting information of the molecular mechanisms responsible for atherosclerosis development (2). Since whole tissue analysis may be very complex, pre-fractionation of the tissue in its layers allows better localization of occurring molecular events. In this way, laser microdissection is the best approach to perform such fractionation, since it constitutes a fast, reproducible, and noncontact isolation methodology (3). The combination of LMD with proteomics has the difficulty of a relatively poor protein yield after LMD, since regions excised are thin and of limited surface area. The sensitivity of mass spectrometers has exponentially increased in the last years (4, 5), allowing the analysis of scarce samples as LMD extracts. In addition,

Fernando Vivanco (ed.), *Vascular Proteomics: Methods and Protocols*, Methods in Molecular Biology, vol. 1000, DOI 10.1007/978-1-62703-405-0_2, © Springer Science+Business Media New York 2013

combination of LMD and protein microarrays has been set up successfully and offers interesting possibilities in the field of differential proteomic analysis (6). Moreover, differential analysis based on 2-DE methodology has overcome this limitation by means of saturation labeling of cysteine residues with fluorochromes (7), which very significantly augments fluorescence of the proteins and provides high-resolution spot maps with very scarce total protein amounts.

Previous studies on coronary artery layers have focused on describing the proteome of the three composing layers of the artery by means of LC-MS/MS (8). We have developed a methodology to isolate such layers by laser microdissection and pressure catapulting (LMPC) and compare them with healthy arteries by means of 2D-DIGE. To date, we have provided the first 2-DE spot maps of arterial intima and media layers (9), and the one 2D-DIGE analysis of intima layer of atherosclerotic coronary, where the pathology mostly develops, revealing altered proteins implicated in the migrative capacity of vascular smooth muscle cells (VSMCs), extracellular matrix (ECM) composition, coagulation, apoptosis, heat shock response, and intraplaque hemorrhage deposition (10).

2 Materials

Aqueous and partially aqueous solutions used during the staining procedure for LMD have to be supplemented with 0.01 % protease inhibitor cocktail (Sigma-Aldrich) and precooled at 4 °C, unless otherwise specified. DTT (BioRad) and Pharmalytes pH 3–10 or pH 4–7 (GE Healthcare) should be added in the moment of using sample and rehydration buffers.

2.1 Histology

1. Optimal Cutting Temperature compound (OCT, Sakura Finetek).
2. Antibodies: Anti-smooth muscle actin antibody, Clone 1A4 (Dako), anti-CD68 antibody, Clone PG-M1 (Dako), REAL Antibody Diluent (Dako), secondary antibody: peroxidase-conjugated EnVision + Dual Link (ready-to-use solution, Dako).
3. Mayer's hematoxylin solution (Sigma-Aldrich), Eosin Y alcoholic solution (Sigma-Aldrich), 70 %, 95 %, and 100 % ethanol, xylene.
4. DPX mounting medium for microscopy (Merck).
5. Formaldehyde solution for molecular biology, 36.5 % (Sigma-Aldrich).
6. Oil Red stock solution: 1 g Oil Red O (Sigma-Aldrich), add 2-propanol (Sigma-Aldrich) up to 100 ml.
7. Glycerol gelatin (Sigma-Aldrich).

8. Hydrogen peroxide 3 %, from hydrogen peroxide 33 %.
9. Wash buffer 1×: From wash buffer 10×, Dako.
10. Blocking solution: BSA 10 % in wash buffer 1×.
11. Liquid DAB+ substrate-chromogen system (Dako).

2.2 Tissue Processing and LMD Isolation

1. Saline solution: Sodium chloride 0.9 % (Braun).
2. Optimal Cutting Temperature compound (OCT, Sakura Finetek).
3. Polyethylene naphthalate (PEN) membrane slides (PALM Microlaser, Carl Zeiss).
4. 70 % and 100 % ethanol (Merck).
5. Certistain Cresyl violet (Merck).
6. Microbeam system (PALM Microlaser, Carl Zeiss).
7. 500 μl opaque adhesive cap tube (PALM Microlaser, Carl Zeiss).
8. Lysis buffer: 7 M urea, 2 M thiourea, 4 % CHAPS, 30 mM Tris, 1 % DTT (*see* **Note 1**).
9. Protein Desalting Spin Column (Pierce).
10. Acetone HPLC-grade (Scharlau).

2.3 Saturation Labeling DIGE Buffers

1. Labeling buffer: 7 M urea, 2 M thiourea, 4 % CHAPS, 30 mM Tris, pH 8.0.
2. pH Fix 7.5–9.5 indicator strips (Macherey-Nagel).
3. 50 mM sodium hydroxide (NaOH) solution, 50 mM hydrochloric acid (HCl) solution.
4. 24 cm IPG strips pH 4–7 (GE Healthcare).
5. Rehydration buffer: 7 M urea, 2 M thiourea, 4 % CHAPS, 1 % Pharmalytes pH 4–7 and 13 mM DTT.
6. 2 mM TCEP solution (Sigma-Aldrich).
7. CyDye DIGE Fluor Labelling Kit for Scarce Samples (GE Healthcare).
8. 2× sample buffer: 7 M urea, 2 M thiourea, 4 % CHAPS, 2 % Pharmalytes pH 3–10 and 130 mM DTT.
9. Protean IEF Cell (BioRad), Ettan Dalt Six (GE Healthcare), 9400 Typhoon Scanner (GE Healthcare).
10. Equilibration buffer: 6 M urea, 50 mM Tris, 30 % glycerol, 2% SDS, pH 8.8.
11. DeCyder 2D Differential Analysis Software v. 7.0 (GE Healthcare).
12. 20 mM TCEP solution.
13. Fixation solution: 30 % ethanol, 5 % acetic acid in bidistilled water.

14. Silver Staining Kit, Protein (GE Healthcare).
15. DP protein digestion station (Bruker-Daltonics).
16. Ammonium bicarbonate (Sigma-Aldrich) 50 mM and 20 mM.
17. 50 % methanol (Sigma-Aldrich), 15 % 2-propanol, 60 % and 30 % acetonitrile, 0.1 % trifluoroacetic acid.
18. 20 ng/μl porcine trypsin (Promega), α-cyano-4-hydroxycinnamic acid (Sigma-Aldrich).
19. 384 Opti-TOF 123×81 mm MALDI plate (AB Sciex), 4800 Plus MALDI TOF/TOF Analyzer (AB Sciex).

3 Methods

3.1 Arterial Sample Collection and Histology (Fig. 1)

1. Wash the tissue with saline at least three times or until blood contamination is eliminated. Dry the specimen with 100 % cellulose paper and place it on an adequate cast for embedding. Add OCT and freeze it with liquid nitrogen. Store it at −80 °C until use (*see* **Note 2**).
2. Cut 5 μm sections, consecutive to the ones to be used for LMD, with a cryostat for histological analysis.
3. Hematoxylin and eosin staining: Fix in cold acetone (−20 °C) for 5 min, remove OCT with tap water, stain in hematoxylin solution for 10 min, wash in running tap water, dip two to three times in eosin solution, dehydrate in ethanol 70 %, 95 %, 100 % solutions, clear in xylol, and mount in DPX.
4. Oil Red O staining: Fixate tissue with 4 % formaldehyde in PBS. Prepare working solution by filtrating 12 ml of stock solution and adding 8 ml of distilled water. Stain slides with working solution for 20 min (*see* **Note 3**). Rinse with distilled water. Stain with hematoxylin for 1 min. Warm glycerol gelatin at 50 °C to melt it and mount the slides (*see* **Note 4**).
5. Actin immunohistochemistry: Fix in cold acetone (−20 °C) for 5 min, remove OCT with tap water, mark selected area with ImmunoPen, incubate with blocking solution for 1 h, followed by smooth muscle actin antibody 1:500 in antibody diluent for 30 min, wash with wash buffer, eliminate peroxidases activity with hydrogen peroxide 3 % for 5 min, wash with wash buffer, incubate with secondary antibody for 30 min, wash with wash buffer, incubate with DAB (5–10 min), wash with water (*see* **Note 5**), stain in hematoxylin solution (30 s to 1 min), wash in running tap water, dehydrate in ethanol 70 %, 95 %, 100 % solutions, clear in xylol, and mount in DPX.
6. CD68 immunohistochemistry: Follow actin staining protocol with anti-CD68 as primary antibody.

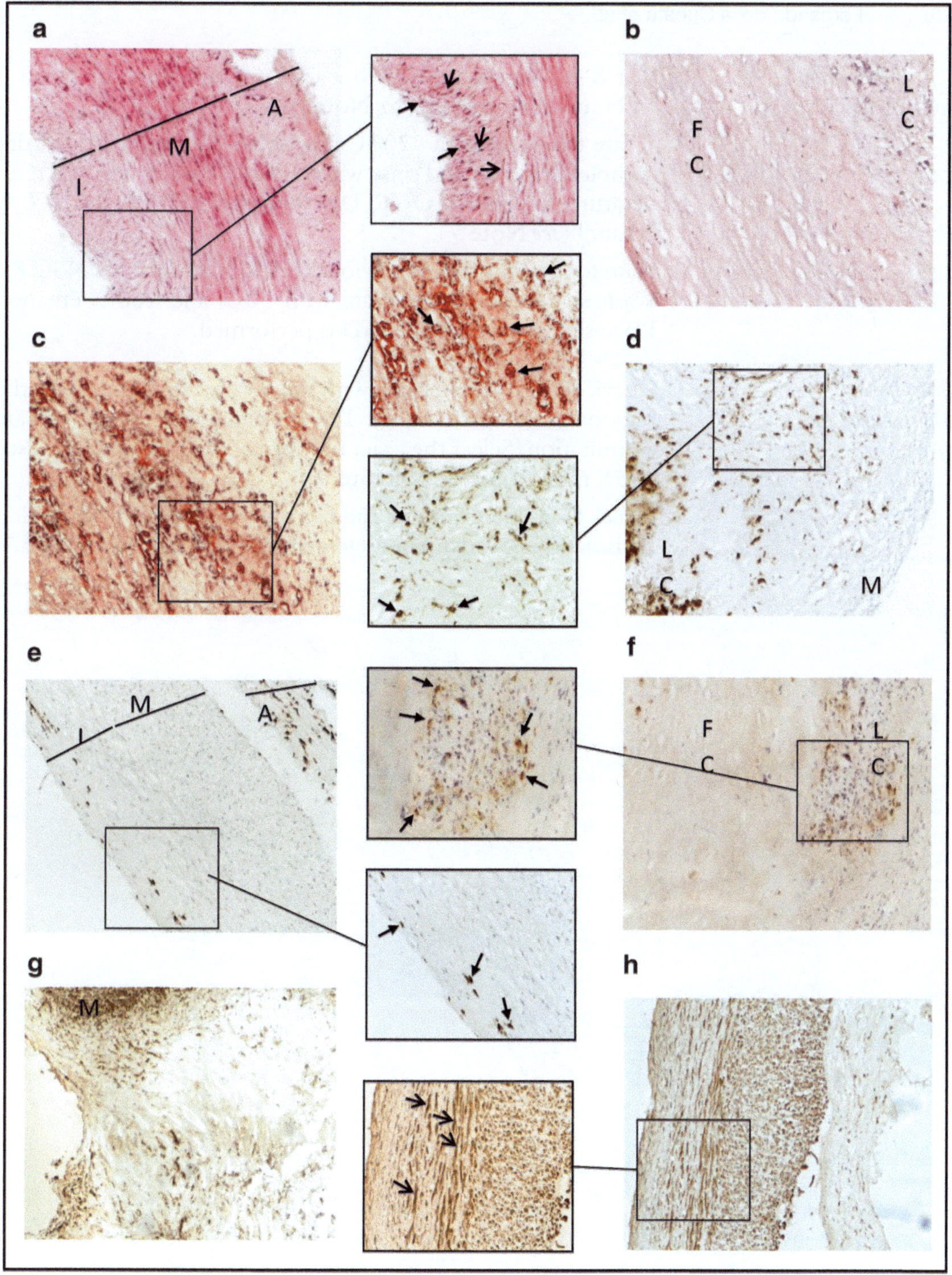

Fig. 1 Histology of atherosclerotic and preatherosclerotic arteries. H&E and Oil Red staining, together with CD68 and actin IHC, were performed with every artery studied in order to identify the artery architecture and to characterize their atherosclerotic lesion degree. (**a**) Preatherosclerotic radial biopsy, H&E. (**b**) Atherosclerotic coronary autopsy, H&E. (**c**) Atherosclerotic coronary autopsy, Oil Red. (**d**) Atherosclerotic coronary biopsy, CD68. (**e**) Preatherosclerotic coronary autopsy, CD68. (**f**) Atherosclerotic coronary autopsy, CD68. (**g**) Atherosclerotic coronary biopsy, actin. (**h**) Preatherosclerotic coronary autopsy, actin. In some images, a subregion has been augmented (200× magnification). *I* intima, *M* media, *A* adventitia, *LC* lipid core, *FB* fibrous cap. *Open-end arrow*: VSMCs. *Closed-end arrow*: macrophages/foam cells. From (10) with permission

3.2 Processing for LMD Isolation

1. Cut 8–10 μm sections with a cryostat and place them on 3 PEN membrane slides (*see* **Note 6**).
2. Fixate the slides with −20 °C precooled 70 % ethanol. Let dry completely on ice and rinse with bidistilled water at room temperature to remove OCT. Dehydrate with 70 % and 100 % ethanol (*see* **Note 7**).
3. Stain for 1 min with Cresyl violet in 100 % ethanol (*see* **Note 8**). Wash staining and dehydrate with 70 % and 100 % ethanol. Place slides on ice until LMD is performed.

3.3 Laser Microdissection (Fig. 2)

1. Check arterial architecture with the 5× objective and select the regions of interest with the 10× objective for a more accurate delimitation. Select the exact number of regions needed to isolate 8 mm^2 of tissue (*see* **Note 9**).
2. Check laser energy power and focus for cutting and catapulting by isolating 2–3 delimited elements before starting automatic

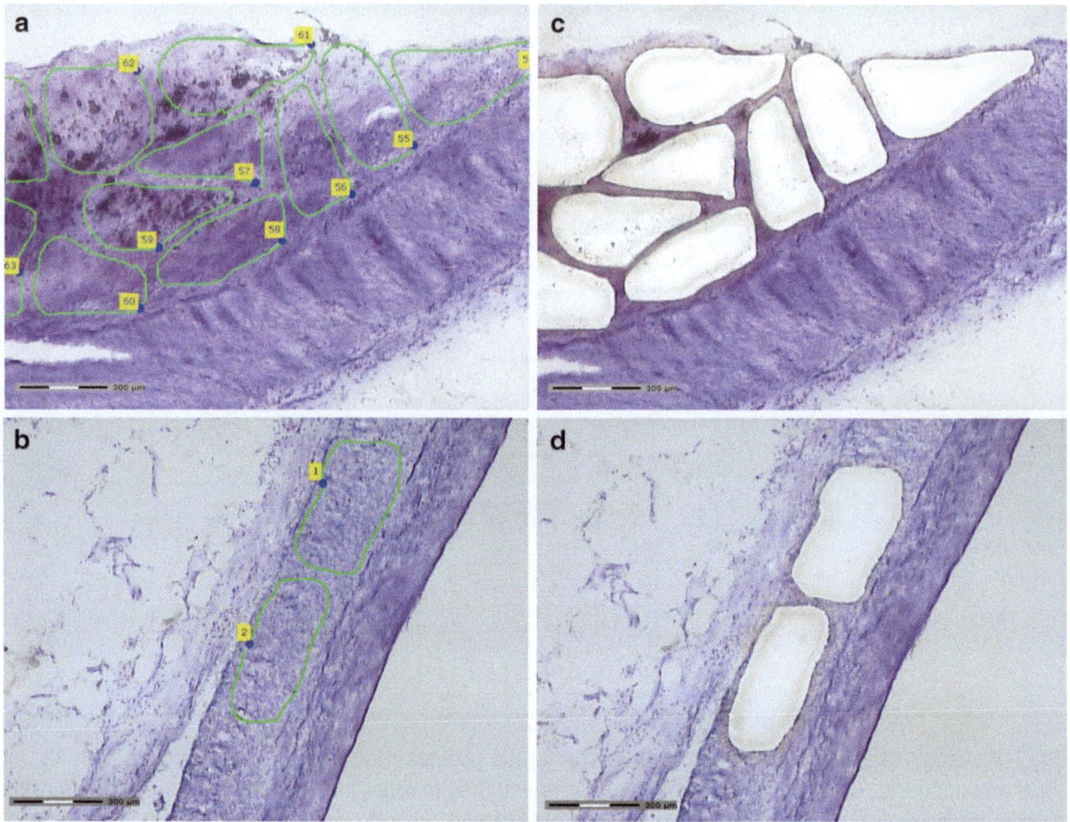

Fig. 2 Isolation of intima and media layers by LMPC. Delimited regions from atherosclerotic coronary intima (**a**) are efficiently isolated by LMD (**b**). Two elements selected within media layer of a preatherosclerotic coronary (**c**) are isolated by LMD, leaving the corresponding blank areas inside the tissue (**d**). From (9) and (10) with permission

cutting function to ensure an efficient isolation of the selected areas. Laser energy power mostly used for arterial regions isolation was 75 % for cutting and 100 % for catapulting. When laser focus and energy have been settled, RoboLPC function can be applied in order to automatically isolate all defined elements (*see* **Note 10**).

3.4 Protein Extraction

1. Collect isolated areas on a 500 μl adhesive cap tube during LMD. Add 100 μl lysis buffer to the tube and incubate with the tube inverted for 5 min on ice. Sonicate three times for 1 min alternated with 1 min on ice. Centrifuge at 12,000 × *g* and place supernatant on a new tube.
2. Place a Protein Desalting Spin Column in a 1.5 ml tube and centrifuge 1 min at 1,500 × *g* to remove retained liquid. Replace the tube for a new one and add the lysate to the column. Centrifuge 2 min at 1,500 × *g* and discard column (*see* **Note 11**).
3. Precipitate the lysate with cold acetone and suspend the pellet in 18 μl labeling buffer.

3.5 Saturation Labeling DIGE (Fig. 3)

1. Check the pH of the lysates to be between 7.8 and 8.2 (ideally 8.0) by spotting 0.2 μl on a pH indicator strip. If necessary, adjust pH by adding either 50 mM NaOH to increase it or 50 mM HCl to lower it (*see* **Note 12**).
2. Rehydrate 24 cm IPG strips pH 4–7 with 450 μl rehydration buffer, 10–24 h previous to sample loading.
3. Create the DIGE internal standard by mixing half of each sample (9 μl).
4. Reduce samples with 1 nmol TCEP (0.5 μl of 2 mM TCEP solution) and the internal standard with the correspondent amount (0.5 μl × number of samples), mix by pipetting, spin down, and incubate 1 h at 37 °C in the dark.
5. Add 2 nmol Cy5 (1 μl of 2 mM resuspended CyDye in DMF) and the correspondent amount to the internal standard (1 μl × number of samples), mix by pipetting, spin down, and incubate 30 min at 37 °C in the dark.
6. Add 1 volume (10.5 μl) of 2× sample buffer.
7. Apply the sample to the rehydrated IPG strips by cup loading. Use the following program for isoelectric focusing (IEF): 200 V for 1 h 30 min, 500 V for 1 h 30 min, 1,000 V for 1 h 30 min, from 1,000 to 8,000 V in 3 h, and 8,000 V until 60,000 V are accumulated.
8. Incubate the IPG strips with 0.01 % DTT in 5 ml of equilibration buffer, 20 min with agitation (*see* **Note 13**).

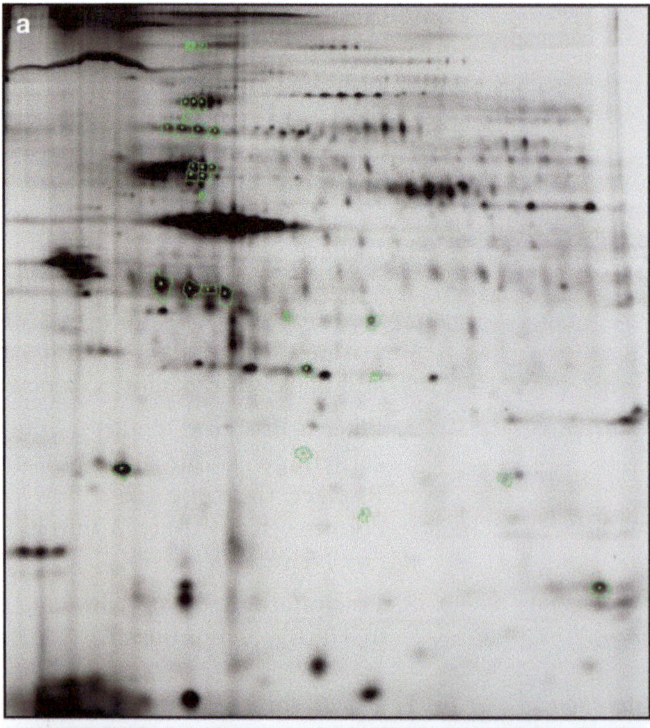

Fig. 3 Representative atherosclerotic coronary 2D-DIGE spot map with differentially expressed spots with respect to healthy coronary and radial artery (*p*-value <0.05, fold change greater than 2.0 or lower than −2.0, Cy5 fluorochrome gray scale representation, from (10) with permission) (**a**). 2D-DIGE spot map of coronary artery media layer (Cy3/Cy5 merged representation) (**b**)

9. Wash the strips with bidistilled water and place them in 12.5 % polyacrylamide gels. Run the gels on the Ettan Dalt Six device (*see* **Note 14**).
10. Scan the gels on a Typhoon Scanner, firstly at 1,000 ppm to adjust laser power and at 100 ppm to get high-resolution images.
11. Perform differential analysis with DeCyder 2D Software.

3.6 Preparative Gels and MS/MS Identification

1. For the identification of differential spots, use a pool of several LMD protein extracts, if you can collect enough material, or otherwise select a reference proteome, which may be a whole artery extract. Reduce 300 μg of protein with 60 nmol (3 μl of 20 mM TCEP solution) TCEP for 1 h at 37 °C in the dark and label with 120 nmol Cy3 dye (6 μl of 20 mM Cy3 resuspended in DMF) for 30 min at 37 °C in the dark (*see* **Note 15**).
2. Add 1 volume of 2× sample buffer and rehydration buffer to reach final volume of 450 μl. Load sample on a 24 cm IPG strip pH 4–7 (*see* **Note 16**).
3. Subject the strip to the same IEF conditions applied for the comparative analysis, except cup loading. Apply in-gel active rehydration at 50 V for 12 h (*see* **Note 17**).
4. Stain the gels with the Silver Staining Kit. Briefly, fixate them with fixation solution for 30 min, incubate in sensitizing solution for 20 min, rinse three times with bidistilled water for 5 min, incubate in 0.25 % silver nitrate solution for 20 min, and incubate with developing solution until the gel staining is adequate, then stop reaction with stopping solution for 10 min and place the gel in bidistilled water (*see* **Note 18**).
5. Excise the differential spots from the gel and digest them in the DP protein digestion station using the protocol of Shevchenko et al. (11) with minor variations. Briefly, rinse with 50 mM ammonium bicarbonate in 50 % methanol and acetonitrile 70 % and dry in a SpeedVac. Add modified porcine trypsin at a final concentration of 20 ng/μl in 20 mM ammonium bicarbonate and proceed at 37 °C overnight.
6. Extract peptides from gel pieces with 60 % acetonitrile and 0.1 % formic acid.
7. Mix an aliquot of the digestion solution with an aliquot of α-cyano-4-hydroxycinnamic acid in 30 % aqueous acetonitrile, 15 % 2-propanol, and 0.1 % trifluoroacetic acid.
8. Deposit using the thin-layer method onto the MALDI plate (Applied Biosystems) and allow drying at room temperature.
9. Obtain MALDI-MS and MS/MS data in an automated analysis loop with the MALDI TOF/TOF Analyzer (*see* **Note 19**).

4 Notes

1. Five different lysis buffers were analyzed and the composition here reported was the best in terms of protein solubilization and compatibility with IEF analysis (9).
2. Do not place the specimen directly in the liquid nitrogen for embedding, since temperature change after freezing may provoke rupture of the material. Place the material inside an extensive polystyrene recipient for a gradual freezing, which will avoid fragmentation.
3. Oil Red crystallizes very easily, depositing in the slide. To avoid these deposits, place slides upside down in a Petri dish and elevate them from the bottom of the dish using any surface in order to allow tissue contact with the staining solution.
4. Do not dehydrate with graded ethanol series, nor clear with xylene, since this would result in lipids solubilization in such organic solvents, removing them from the tissue. Glycerol gelatin is an aqueous mounting medium and has to be used in this case instead of a nonaqueous mounting medium, like DPX.
5. Prepare DAB solution using gloves, carefully remove it after incubation, and render it inactive with sodium hypochlorite before wasting it in an adequate biohazard container, since it may cause cancer after longtime exposure.
6. Thicker sections would result on greater protein recovery, but thinner sections may be necessary to isolate smaller regions than whole layers. In the latter, 10 μm allow to very specifically isolate them providing enough protein material for subsequent proteomic analysis. PEN membrane slides permit greater regions' catapulting which relies on greater protein recovery and allows checking efficient collection of the catapulted areas by orienting the microscope through the tube cap.
7. Removal of OCT with 4 °C precooled bidistilled water removes less efficiently the compound, since it becomes liquid at RT. Remaining OCT alters or even may impede laser cutting of the tissue.
8. Mayer's hematoxylin provided similar results than Cresyl violet but rendered weaken staining and cannot be solved in pure ethanol, which implies greater proteases activation, which may be important with LMD extended periods. Do not mount and/or cover slides since this would impede LMD. Slides can be stored at -80 °C on a sealed recipient and thawed for LMD isolation when necessary, but this has to be done inside the sealed recipient, since condensation events during thawing may result in tissue hydration that would alter laser cutting. Slides on ice have to be tempered for 5 min before starting LMD.

9. The use of a light diffuser deeply improves visualization of non-mounted specimens, which otherwise are seen in a dark artifact coloring, especially if choosing hematoxylin for staining. Selected regions should not exceed 0.2 mm^2 size, since they may not be well catapulted to the collection tube cap. Although 8 mm^2 provided an adequate protein content for the subsequent 2D-DIGE analysis, greater sample collection may be possible within an adequate LMD timing (<2 h), depending on the efficiency of the procedure and the sample.
10. After applying automatic RoboLPC function, screen the whole slide for areas not well catapulted or that may have fallen from the tube, and use LPC function to manually catapult them back to the tube cap. These mishitting catapulting events are more common at the final steps of the process, where the cap is almost full of material.
11. Tissue staining deeply affects protein solubilization and fluorescent DIGE signal acquisition. Protein Desalting Spin Columns have shown a very efficient removal of remaining staining dye, which in contrast could not be eliminated by acetone precipitation.
12. In order to avoid protein loss during pH checking, pH variation after lysis of the studied sample should be assayed with a preliminary LMD extract and tested during the DIGE experiment with just a few samples to corroborate.
13. In contrast with conventional 2-DE and minimal labeling 2D-DIGE experiments, there is no need to incubate the strips with IAA since cysteine residues will be bound to the saturation labeling CyDye.
14. Run the gels applying constant power, firstly 1 W for all the gels, and when the front dye has clearly entered the gel, apply 20 W per gel for an overday run or 1 W per gel for an overnight one.
15. MS/MS identifications performed in parallel using Cy3 labeled preparative gels and non-labeled ones showed no labeling alterations of the spot pattern in this type of sample, since 29 out of 30 differential spots were identically identified, and the one left could be identified in the non-labeled gel but not in the labeled one.
16. Rehydration buffer for active in-gel rehydration should carry pH 3–10 Pharmalytes, since in this case, protein sample is loaded during rehydration and they facilitate solubilization of all proteins in this pH range, including proteins between 3–4 and 7–10 pH range, which will be excluded with 4–7 pH range Pharmalytes.
17. Cup loading may result on protein precipitation within the cup while working with high protein amounts.

18. Avoid glutaraldehyde in the sensitizing solution since it induces cross-linking of proteins, interfering with mass spectrometry.
19. Analysis of mass data may be performed using the 4000 Series Explorer Software (Applied Biosystems). MALDI-MS and MS/MS data are combined through the GPS Explorer Software to search a nonredundant protein database, such as Swiss-Prot or nrNCBI. Include Cy3 modification in the search engine fixed modifications box.

Acknowledgments

This work was supported by grants from Instituto de Salud Carlos III (PI11/02239, FIS PI070537, FIS PI080970) and Mutua Madrileña Automovilista (20174/004), Redes temáticas de Investigación Cooperativa en Salud (RD06/0014/1015), and Fundación para la Investigación Sanitaria de Castilla-La Mancha (FISCAM PI2008-08, PI2008-28, PI2008-52).

References

1. Hanson GK (2005) Inflammation, atherosclerosis and coronary artery disease. N Engl J Med 352:1685–1695
2. De la Cuesta F, Alvarez-Llamas G, Gil-Dones F, Martin-Rojas T, Zubiri I, Pastor C et al (2009) Tissue proteomics in atherosclerosis: elucidating the molecular mechanisms of cardiovascular diseases. Expert Rev Proteomics 6:395–409
3. Burgemeister R (2005) New aspect of laser microdissection in research and routine. J Histochem Cytochem 53:409–412
4. Domon B, Aebersold R (2006) Mass spectrometry and protein analysis. Science 312: 212–217
5. Ahrens CH, Brunner E, Qeli E, Basler K, Aebersold R (2010) Generating and navigating proteome maps using mass spectrometry. Nat Rev Mol Cell Biol 11:789–801
6. Espina V, Wulfkuhle J, Liotta LA (2009) Application of laser microdissection and reverse-phase protein microarrays to the molecular profiling of cancer signal pathway networks in the tissue microenvironment. Clin Lab Med 29:1–13
7. Greengauz-Roberts O, Stöppler H, Nomura S, Yamaguchi H, Goldenring JR, Podolsky RH et al (2005) Saturation labeling with cysteine-reactive cyanine fluorescent dyes provides increased sensitivity for protein expression profiling of laser-microdissected clinical specimens. Proteomics 5:1746–1757
8. Bagnato C, Thumar J, Mayya V, Hwang SI, Zebroski H, Claffey KP et al (2007) Proteomics analysis of human coronary atherosclerotic plaque: a feasibility study of direct tissue proteomics by liquid chromatography and tandem mass spectrometry. Mol Cell Proteomics 6:1088–1102
9. De la Cuesta F, Alvarez-Llamas G, Maroto AS, Donado A, Juarez-Tosina R, Rodriguez-Padial L et al (2009) An optimum method designed for 2-D DIGE analysis of human arterial intima and media layers isolated by laser microdissection. Proteomics Clin Appl 3:1174–1184
10. De la Cuesta F, Alvarez-Llamas G, Maroto AS, Donado A, Zubiri I, Posada M et al (2011) A proteomic focus on the alterations occurring at the human atherosclerotic coronary intima. Mol Cell Proteomics 10, M110.003517
11. Shevchenko A, Wilm M, Vorm O, Mann M (1996) Mass spectrometric sequencing of proteins silver-stained polyacrylamide gels. Anal Chem 68:850–858

Chapter 3

Use of TOF-SIMS in Vascular Biology

Sebastián Mas, Raúl Pérez, and Jesús Egido

Abstract

Secondary ion mass spectrometry is a technique used for surface analysis. The recent introduction of metal cluster ionization sources opened the possibility to probe biological tissues, due to its reduced in-source fragmentation. Despite its limitations in mass range of the molecules analyzed, a combination of minimal sample processing, sensibility, and unsurpassed lateral resolution makes TOF-SIMS a good tool for molecular imaging in pathology.

Key words Atherosclerosis, Imaging, Lipidomics, Metabolomics, SIMS-TOF

1 Introduction

SIMS imaging has been employed for more than 30 years, mainly is solid-state physics and electronics. Since early 1980s, the advent of charge neutralization (1) made surface probing less aggressive, and it could be employed for organic analysis. But it was until mid-1990s, with the coupling of cluster secondary ion ionization to a time-of-flight (TOF) analyzer, when its use in biological materials became possible.

1.1 SIMS Principles

Secondary ion mass spectrometry (SIMS) as is it configured nowadays analyzes the uppermost layer of a two-dimensional sample. For doing this, in an SIMS experiment, a focused ion beam (primary ions) is used to bombard the surface. Typical energies of the primary ions are in the keV range, so the direct collisions between the primary ions and the atoms in the sample are highly energetic compared to bond energies. This results in extensive fragmentation (destructive technique) and bond breaking near the collision site; some intact molecules can then be emitted from the first layers near the surface if they acquire enough energy to overcome the surface binding energy. In most cases, the molecules are found to desorb from within 5–10 nm of the impact point (2). Most of them

Fernando Vivanco (ed.), *Vascular Proteomics: Methods and Protocols*, Methods in Molecular Biology, vol. 1000, DOI 10.1007/978-1-62703-405-0_3, © Springer Science+Business Media New York 2013

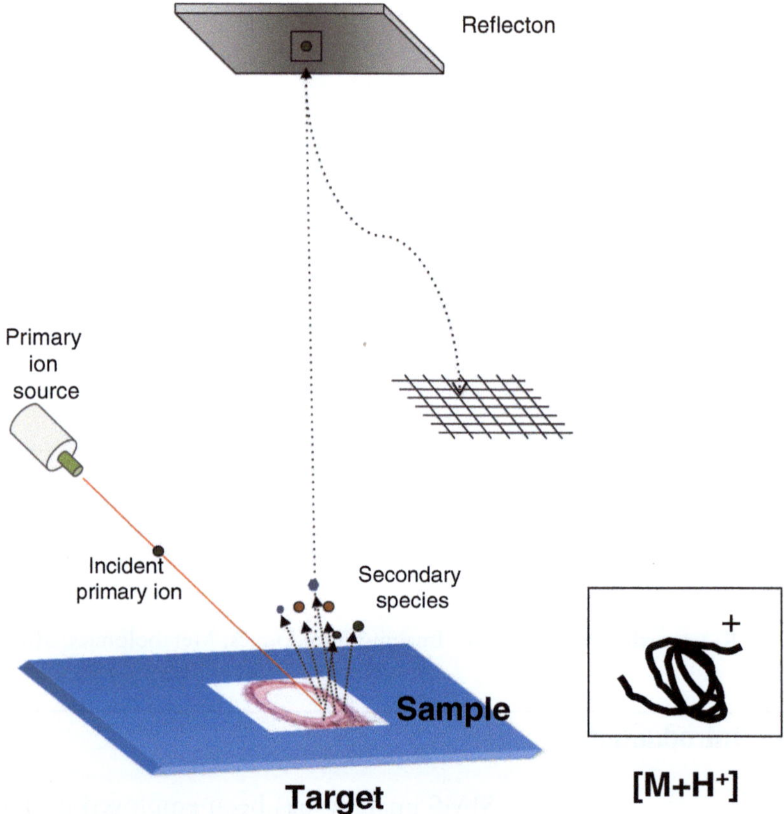

Fig. 1 TOF-SIMS surface rastering and ionization schema

are emitted as neutrals and only a small fraction (1 %) is charged, and their positive or negative state depends on their electron configuration (Fig. 1).

In order to minimize sample damage for biomolecules, the primary ions impinging the surface must be low enough ($<10^{13}$ ions/cm^2) to ensure that every area is not impacted more than once, which is called static SIMS regime (3).

1.2 TOF-SIMS Instrumentation

As in any other mass spectrometer, types of instrumentation depend on ion sources (ion guns in this case) and analyzer employed. Although in biological analysis, all equipment employed is fitted with TOF analyzer. In such configuration a pulsed primary ion gun is used to generate secondary ions (4, 5).

Secondary ions emitted from the tissue surface are extracted into a TOF analyzer by a high-voltage potential, and their mass is determined by measuring the time that takes the ions to reach the detector (time of flight). Flight time is converted via calibration into mass, provided flight distance and extraction energy remain constant, although observed mass is proportional to ion charge (z).

The mass spectra appear as an *m/z* vs. intensity plot of all ions present in the sample. The accuracy in the measurements improves as the in-pulse shortens, so subnanosecond primary ion pulses render a minimal energy spread affecting dramatically the resolution achieved at the detectors.

In a TOF detector secondary ions of all masses are collected in parallel, and the spectrum is accumulated from many primary pulses. This feature of TOF-SIMS together with the high transmission of the analyzer and the low background due to the single-ion counting allows very high sensitivity and accuracy for surface species.

1.3 TOF-SIMS as a Tool in Pathology

The use of TOF-SIMS in pathology is limited due to the extensive fragmentation of large biomolecules; thus, the amount of molecular information obtained from the surface is practically limited to metabolites (<1,000 Da).

Nevertheless, its combination of sensitivity, lateral resolution, and minimal sample preparation makes this technique very attractive in biomedical research. Most of the studies published in the last five years in the literature are focused on quantitation and distribution of lipids. Lipid imaging research is due to two confluent factors: first of all, technical factors as cluster liquid metal ion guns showed a strong bias towards hydrophobic molecules with sensitivities as low as attomolar. Secondly, lipidomics has attracted a lot of attention in pathology due to the description of an increasing number of bioactive lipids and their involvement in human pathology.

Thus, the lipid imaging research is focused mainly in two areas: first, subcellular distribution of lipids during the cell cycle in cultures. Thus, Winograd and coworkers pioneer this research using Tetrahymena as model to unicellular organisms mating (6, 7). Neurons have also been employed to probe the lipid distribution within the cell in Aplysia (8) or mouse ganglia (9) neuron cultures.

The other active research line and focus of this chapter is the use of TOF-SIMS to imaging biopsies of lipid accumulation pathologies as lysosomal storage diseases (i.e., Fabry disease (10)), steatosis (i.e., nonalcoholic fatty liver disease (11)), or atherosclerotic diseases (12–14).

2 Materials

2.1 Sample Preparation

1. Liquid nitrogen container.
2. Cold-resistant beaker.
3. 2-methylpentane (Sigma, M32631).
4. Tissue-Tek OCT Compound (Optimal media cutting) (Sakura).

5. Cryostat blades and chucks (Leica).
6. Cryostat (CM1950, Leica).
7. Stainless steel plate (15-7PD, Goodfellow).
8. IHC FLEX Microscope Slides (Dako).
9. Vacuum lyophilizer (freeze dryer) or vacuum desiccator.

2.2 TOF-SIMS Analysis

1. Planar aluminum holder stage to support glass slide with sample prepared.
2. Small screws and clamps to hold the glass slide to the holder stage.
3. ION-TOF IV (ION-TOF GmbH, Münster, Germany) TOF-SIMS system equipped with bismuth LMIG.
4. Bi_3^{++} ion selection for maximum yield on high masses.
5. Electron flood gun, W filament, 20 eV electron for charge compensation.
6. Software IonSpec 4.1 and Ion Image 3.1 from ION-TOF GmbH.

3 Methods

3.1 Sample Preparation

TOF-SIMS as any other mass spectrometer performs the analysis under high vacuum, so we cannot perform analysis of living cell or tissues. Sample preparation, in order to preserve all the molecular features, is mandatory and becomes the most critical step.

Tissue fixation cannot be performed using conventional histology protocols as lipids and metabolites diffuse on organic solvents. Using cryogenic storage and processing preserve in vivo ion distribution, avoiding loss or redistribution of easily diffusible ions without any alteration in the morphology. Thus, the biopsies should be immediately processed after the surgical removal, but it is advisable to cut the sample on an appropriate size and orientation with a scalpel prior to freeze it.

3.1.1 Sample Preservation

Tissue is snap frozen by quenching in liquid nitrogen; alternatively you can use supercooled isopentane (2-methyl butane) to prevent ice crystal formation within the tissue (Leidenfrost phenomenon). For this purpose place a cold-resistant plastic beaker containing isopentane into the liquid nitrogen and wait until the isopentane is almost solidifying, and submerge the sample until frozen. Samples are stored at a minimum of −70° until their use.

3.1.2 Cutting

The surface morphology of the sample can affect the generation of secondary ions. Samples with surfaces as flat as possible are desired in order to avoid possible shadow effects; to achieve this, tissues had to be cut in a cryostat: without defrosting biopsies are attached to a cryostat chuck with a small amount of OCT compound (OCT contact should be minimized, as polymeric composition constitutes a contaminant in mass spectrometry). The sample is then put into the cryostat where it must be left to warm to cryostat temperature before cutting; 15–20 min is usually adequate. Check that the temperature is at −25 °C for fatty-acid-rich tissues. For softer tissues you may need to raise the temperature. Harder tissues produce better sections at lower temperatures and vice versa. Adjust the section thickness using the micrometer screw to 6–8 μM, but you may have to increase this for difficult tissues. Trim the tissue block until the wanted section avoiding OCT-embedded parts. Clean the anti-roll and replace the knife if they are in contact with OCT to prevent contaminations. Collect the tissue section with a clean brush and deposit onto a stainless steel plate. Alternatively, microscopy slides can be used despite they are electrical insulators, without major alteration in the resolution (Fig. 2).

3.1.3 Surface Washing

The cleanliness of the sample surface is also an important consideration in sample preparation, as any contaminant adsorbed on the surface during handling or processing can affect the accuracy of the analysis. But in our hands, the only contaminant found is due to OCT polymers. If aforementioned procedures are followed, there is no need of any additional washing.

3.1.4 Dehydration

As it has been mentioned, sample must remain stable under UHV conditions in the SIMS analysis chamber. In order to avoid interferences in ion transmission within the instrumentation, all volatiles including water should be removed prior to its analysis.

Freeze-drying the sample using a vacuum lyophilizer for few minutes has proven to be the best method to prevent the sample thaw while analyzing. Alternatively a simple vacuum desiccator can be employed placing the plates or the slides on top of a precooled blue ice block.

3.2 TOF-SIMS Measurements

The methods described are specific for ION-TOF IV TOF-SIMS instrumentation. The TOF-SIMS analyses are performed in an ultra-high vacuum chamber operated at a pressure of 5×10^{-9} mbar. Samples are bombarded with a pulsed bismuth liquid metal ion source (Bi_3^{++}), at energy of 25 keV. The gun is operated with a 20 ns pulse width, 0.3 pA pulsed ion current for a dosage lower than 5×10^{11} ions/cm^2, well below the threshold level of 1×10^{13} ions/cm^2 generally accepted for static SIMS conditions. Secondary ions are detected with a reflector TOF analyzer, a multichannel plate (MCPs), and a time-to-digital converter (TDC). Measurements are performed with a typical acquisition time of 10 s, at a TDC time resolution of 200 ps and 100 us cycle time.

The cycle time limits the mass range acquisition which typically is established in the 0 to 2.000 amu. Charge neutralization is

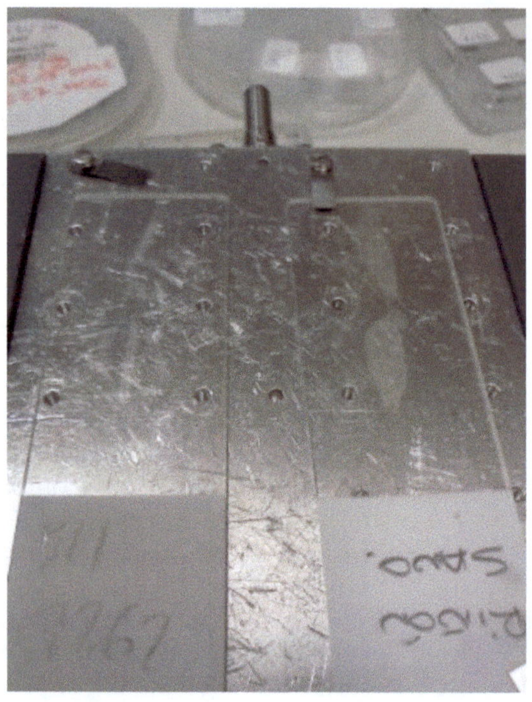

Fig. 2 Biopsies glass slides into a holder prior to its analysis

achieved with a low-energy (20 eV) pulsed electron flood gun. Secondary ions are extracted with 2 kV voltages and are post-accelerated to 10 keV kinetic energy just before hitting the detector. Mass spectral acquisition and calibration are performed within the ION-TOF Ion Spec software (version 4.1). Each spectrum is normalized to the total intensity.

3.3 Data Analysis

3.3.1 Calibration

Due to the very low initial kinetic energy distribution of the secondary ions in SIMS and because of the very high precision of TDCs, the relationship between mass/charge (*m/z*) and the TOF remains linear over the whole mass range (15).

Thus, internal mass calibration can be based on light mass fragment ions that are always present in the spectra, for example, H1, C1, CH1, CH21, and C3H2 1 ion peaks in the positive polarity and C, CH, CH2, C2, C3, and C4H ion peaks in the negative ion mode.

Then, this mass calibration can be refined step by step using identified ions of larger *m/z*. The deduced mass accuracies are in the range of few parts per million (ppm) (16).

3.3.2 Chemical Identification

Each TOF-SIMS spectrum contains an immense number of peaks arising from the fragmentation of the molecular species at the surface, thereby providing a useful fingerprint. By evaluating the masses of the signals, peaks often can be identified from the molecular ion of the analyte, fragments of the molecular ion, and ions of any other components that may be present in the sample (Fig. 3).

Some a priori knowledge of the surface chemical composition can help in this task. This evaluation can be done either by comparison to model compounds or with the aid of library spectra. Furthermore, the latest versions of software packaged with TOF-SIMS instruments contain databases with searchable spectral libraries that allow "peak identification." The exact mass of the peak of interest is considered, and a list of possible ion identities is generated. In spite of these approaches, analysis of TOF-SIMS spectral data can still be overwhelming. Although there is a growing understanding of how the TOF-SIMS spectrum forms, there exists as yet no universal systematic route for data evaluation.

Generally, organic molecules, particularly lipids, are relatively easy to separate from the concomitantly analyzed atomic ions and larger inorganic fragments. Recently, chemometric pattern recognition techniques have been applied to identify and classify samples by their TOF-SIMS spectra (17). These techniques have been widely used in other MS techniques and can be equally useful in TOF-SIMS analysis. Chemometrics pattern recognition methods involve clustering the samples into groups that are related based on their location in multivariate space.

Principal component analysis (PCA) has been the most commonly applied method to TOF-SIMS data spectra (18). This is a statistical method employed to find the combination of variables that describe the most important trends in datasets. PCA allows the classification of similar species and gives indication as to which peaks are contributing to a particular component. This type of analysis allows a fair degree of automation to analyzing a dataset and can allow distinction between spectra, which nominally look very similar.

3.3.3 Imaging

The generation of secondary ion images requires to scan the focused primary ion beam over the sample region of interest via a so-called raster and to record the intensity of selected ion in color-coded intensity map. Each position of this raster is called a pixel. From each of these pixels (typically 256×256 image pixel size), a complete mass spectrum is obtained. So an acquisition contains as many secondary ion images as peaks in the spectra (Fig. 4).

In TOF-SIMS instruments, there are two ways to manage the raster of the sample surface. In the first, the sample is motionless while the primary ion beam moves over it from 1 pixel position to the next. The size of the image that can be acquired with this procedure is limited to the field of the secondary ion extraction optics

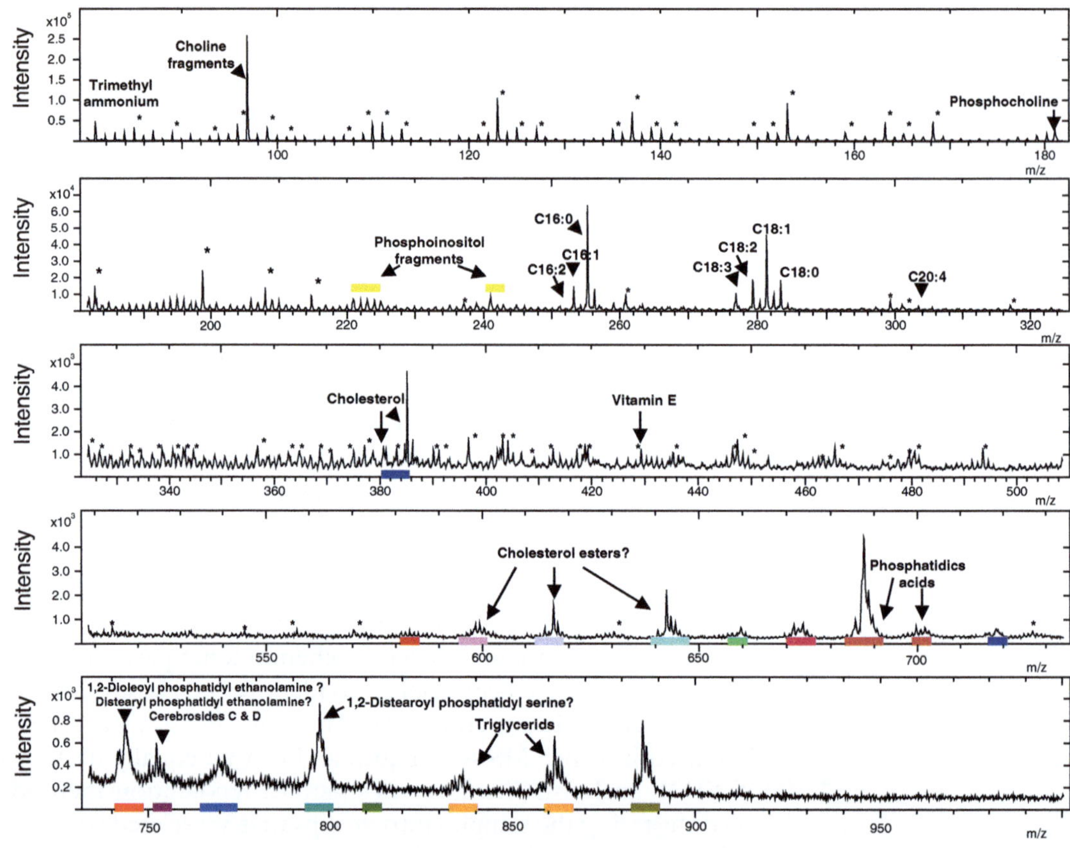

Fig. 3 Example of secondary ion spectra recorded from kidney sample, with putative identification of the most abundant ions

(which is usually within the upper limit of 500×500 mm^2), and the spatial resolution is limited to the beam focus. The second way is applied to samples that require analysis areas greater than the field of the secondary ion extraction optics, which is usually the case of biological/tissue samples. In this case the primary ion beam is not moved but rather the sample holder is moved step by step from 1 pixel to the next pixel position. This method allows the analysis of sample areas having a size of several square centimeters. In this case, the pixel size is generally much larger than the primary ion beam spot, and the beam is rastered inside each pixel in order to avoid under-sampling during data acquisition (19).

Another operating mode can be used, which delivers beam sizes of 100–300 nm (depending on the chosen primary ion species). In this case, the ion beam cannot be bunched and the pulse duration is 20–100 ns, ensuring only a nominal mass resolution. While this may not pose a problem for well-separated peaks with no serious interferences, it can pose a major problem with closely spaced peaks. Moreover, the small beam diameter reduces the

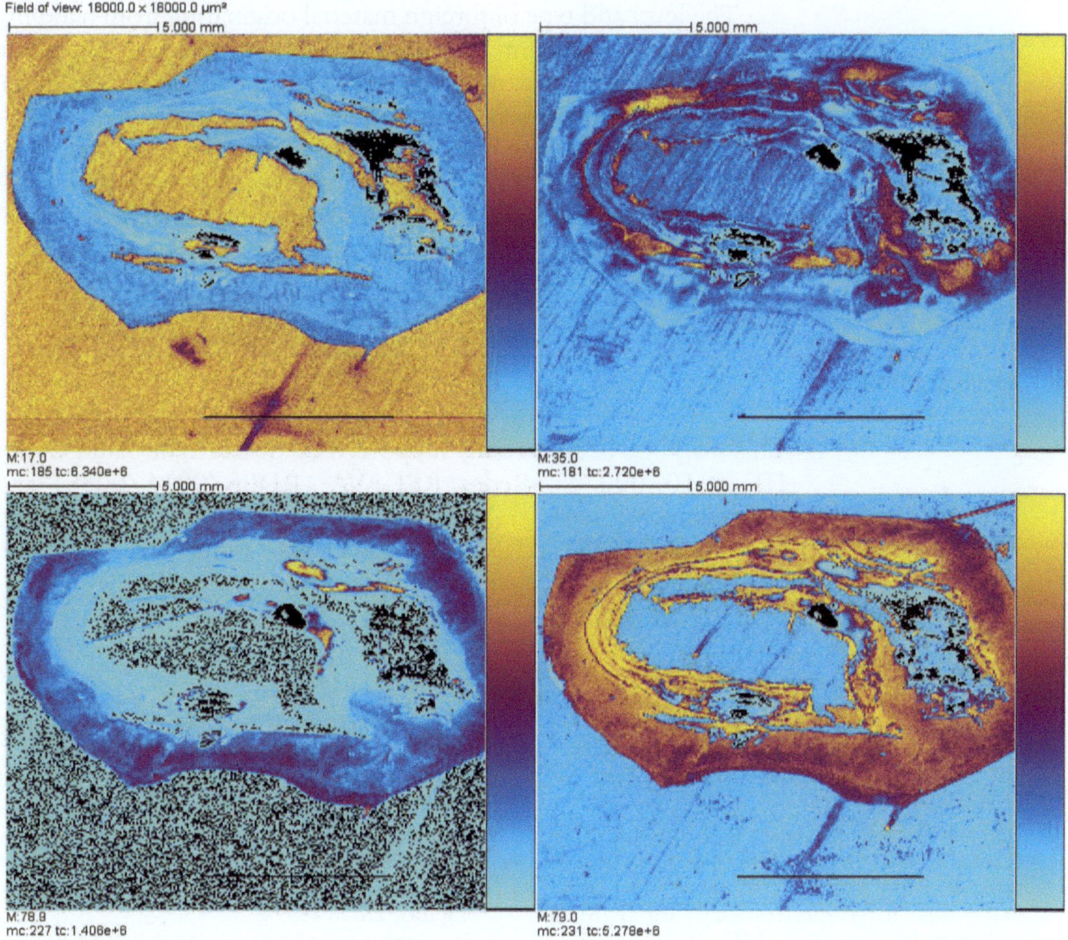

Fig. 4 Representative image of TOF-SIMS in a human biopsy: molecular distribution of several inorganic ions on a human vessel carrying an atherosclerotic plaque

amount of substance available per pixel (image point) in the uppermost monolayer; this limits the number of secondary ions and dynamic range for each point.

However, with TOF-SIMS imaging, a region of interest from the total ion image can be identified and the mass spectra from the pixels in that region can be summed, allowing spectral evaluation with restored sensitivity and dynamic range.

4 Notes

1. The high sensitivity and surface selectivity of TOF-SIMS make it particularly susceptible to inadvertent contamination from sample handling, storage, and mounting. Not accurate handling practices can introduce impurities detectable by TOF-SIMS.

The level and type of foreign material originating from handling may or may not interfere with the analysis at hand, but it is a good practice to avoid procedures that can introduce additional contamination whenever possible. In this sense it is recommended not to use latex gloves when handling the sample, as they are a source of volatile silicones. Instead the use of polyethylene gloves is recommended. Moreover, to keep the sample in a pristine state nothing should come in contact with the sample's surface.

Acknowledgments

This work has been partially supported by Instituto de Salud Carlos III (PI10/00072), Redes RECAVA (RD06/0014/0035) and European Network (HEALTH F2-2008-200647), Euro Salud EUS2005-03565. cvREMOD (091100), Fundacion Lilly and FRIAT.

References

1. Benninghoven A (1973) Surface investigation of solids by the statistical methods of secondary ion mass spectroscopy (SIMS). Surf Sci 35:427–457
2. Bolbach G, Viari A, Galera R, Brunot A, Blais JC et al (1992) Organic film thickness effect in secondary ion mass spectrometry and plasma desorption mass spectrometry. Int J Mass Spectrom Ion Process 112:93–100
3. Vickerman JC (2001) Introduction. In: Vickerman JC, Briggs D (eds) TOFSIMS—an overview. TOF-SIMS—surface analysis by mass spectrometry, Surface Spectra and IM Publications, Manchester and Chichester, pp 1–40
4. Touboul D, Brunelle A, Laprevote O (2006) Structural analysis of secondary ions by post-source decay in time-of-flight secondary ion mass spectrometry. Rapid Commun Mass Spectrom 20:703–709
5. Brunelle A, Laprevote O (2007) Recent advances in biological tissue imaging with time-of-flight secondary ion mass spectrometry: polyatomic ion sources, sample preparation, and applications. Curr Pharm Des 13:3335–3343
6. Ostrowski SG, Van Bell CT, Winograd N, Ewing AG (2004) Mass spectrometric imaging of highly curved membranes during Tetrahymena mating. Science 305:71–73
7. Kurczy ME, Piehowski PD, Van Bell CT, Heien ML, Winograd N, Ewing AG (2010) Mass spectrometry imaging of mating Tetrahymena show that changes in cell morphology regulate lipid domain formation. Proc Natl Acad Sci USA 107(2751):2756
8. Arlt S, Beisiegel U, Kontush A (2002) Lipid peroxidation in neurodegeneration: new insights into Alzheimer's disease. Curr Opin Lipidol 13:289–294
9. Yang HJ, Ishizaki I, Sanada N, Zaima N, Sugiura Y, Yao I, Ikegami K, Setou M (2010) Detection of characteristic distributions of phospholipid head groups and fatty acids on neurite surface by time-of-flight secondary ion mass spectrometry. Med Mol Morphol 43:158–164
10. Touboul D, Roy S, Germain DP, Chaminade P, Brunelle A, Laprevote O (2007) MALDI-TOF and cluster-TOF-SIMS imaging of Fabry disease biomarkers. Int J Mass Spectrom 260:158–165
11. Debois D, Bralet MP, Le Naour F, Brunelle A, Laprevote O (2009) In situ lipidomic analysis of nonalcoholic fatty liver by cluster TOF-SIMS imaging. Anal Chem 81:2823–2831
12. Mas S, Touboul D, Brunelle A, Aragoncillo P, Egido J, Laprevote O, Vivanco F (2007) Lipid cartography of atherosclerotic plaque by cluster-TOF-SIMS imaging. Analyst 132:24–26
13. Malmberg P, Börner K, Chen Y, Friberg P, Hagenhoff B, Månsson JE, Nygren H (2007) Localization of lipids in the aortic wall with imaging TOF-SIMS. Biochim Biophys Acta 1771:85–95

14. Mas S, Martínez-Pinna R, Martín-Ventura JL, Pérez R, Gomez-Garre D, Ortiz A, Fernandez-Cruz A, Vivanco F, Egido J (2010) Local non-esterified fatty acids correlate with inflammation in atheroma plaques of patients with type 2 diabetes. Diabetes 59:1292–1301

15. Postawa Z, Czerwinski B, Winograd N, Garrison BJ (2005) Microscopic insights into the sputtering of thin organic films on Ag{111} induced by C60 and Ga bombardment. J Phys Chem B 109:11973

16. Passarelli MK, Winograd N (2011) Lipid imaging with time-of-flight secondary ion mass spectrometry (Tof-SIMS). Biochim Biophys Acta 1811:976–990

17. Jungnickel H, Jones EA, Lockyer NP, Oliver SG et al (2005) Application of TOF-SIMS with chemometrics to discriminate between four different yeast strains from the species Candida glabrata and Saccharomyces cerevisiae. Anal Chem 77:1740–1745

18. Tyler BJ, Rayal G, Castner DG (2007) Multivariate analysis strategies for processing ToF-SIMS images of biomaterials. Biomaterials 28:2412–2423

19. Todd PJ, Schaaff TG, Chaurand P, Caprioli RM (2001) Organic ion imaging of biological tissue with secondary ion mass spectrometry and matrix-assisted laser desorption/ionization. J Mass Spectrom 36:355–369

Chapter 4

Proteomic Analysis of Vascular Smooth Muscle Cells with S- and R-Enantiomers of Atenolol by iTRAQ and LC-MS/MS

Jianhua Zhang, Jiahua Shi, Benedict Jia Hong Lee, Liwei Chen, Kee Yang Tan, Xiaoling Tang, Jane Yi Lin Tan, Xiang Li, Huixing Feng, and Wei Ning Chen

Abstract

The term "proteome" was first introduced by referring to the complete determination of proteins expressed in a given cell, tissue, or organism. In HPLC-based proteomics technique, the mixture of the cleaved peptides are bonded, separated sequentially on the multidimensional columns based on charge or hydrophobicity of the ionized analytes, and then eluted into the MS for identification. Among the developed stable isotope-based quantification methods, iTRAQ has recently gained popularity as its simple iTRAQ labeling procedures and up to eight labeled samples examined in a single experiment.

Key words iTRAQ, LC-MS/MS, Proteomics, Chiral drug, Cardiac adrenoreceptors, Atenolol

1 Introduction

LC-MS has become the technique of choice for proteomics study for the separation and identification of the polypeptides (1–7). Online libraries of information and sequence structures of polypeptides are now available to aid in efficient identification of the peptide sequences. Published articles summarize and discuss the LC-MS technique development and applications for proteomics. Using C. elegans model, Takahashi et al. showed that most proteins could be detected using the LC-MS technique. The most acidic protein identified has a pI 3.48, while the most basic one has a pI 12.41. The smallest gene product detected has a MW 6.0 kDa and the largest one has a MW 1,369 kDa (8). Shen research group discussed the advances and limitations of LC-MS proteome analysis for biomarker discovery, including quantitative proteomics, deep proteome profiling, and phosphoproteome profiling (9). Another review summarized recent advances in LC-MS-based

Fernando Vivanco (ed.), *Vascular Proteomics: Methods and Protocols*, Methods in Molecular Biology, vol. 1000, DOI 10.1007/978-1-62703-405-0_4, © Springer Science+Business Media New York 2013

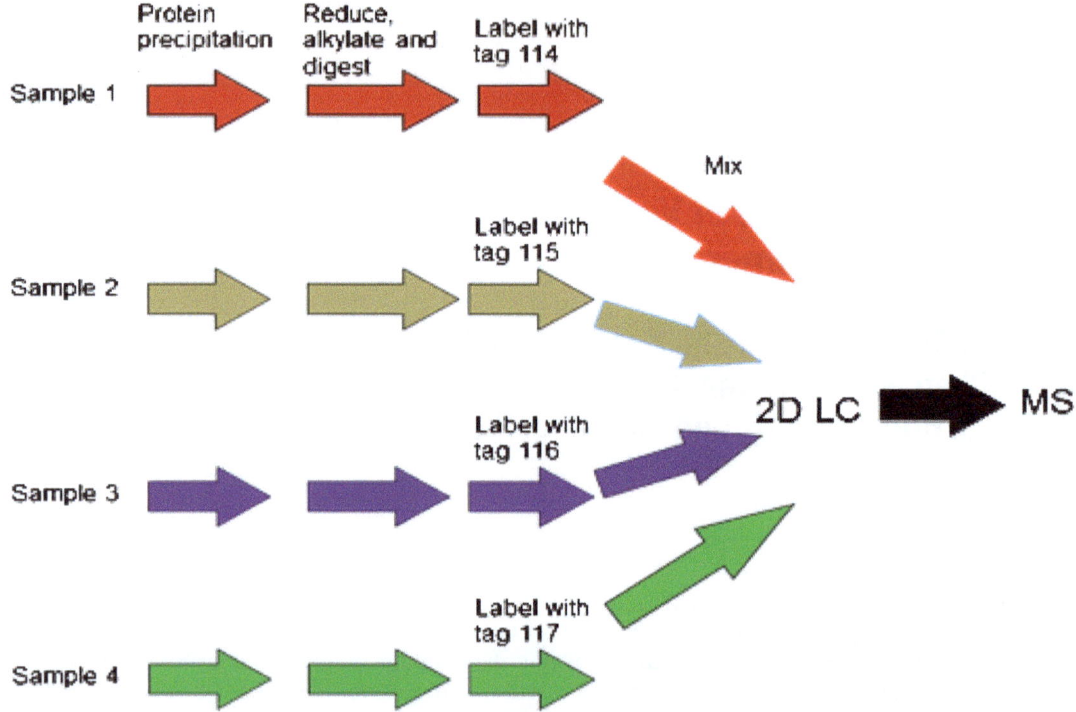

Fig. 1 A typical schematic of workflow for the preparation of the samples labeled with iTRAQ reagents

proteomics profiling and its applications in clinical proteomics as well as discussed the challenges of more effective candidate biomarker discovery using these technologies (6). In a review, Tannu et al. examined techniques using two-dimensional gel electrophoresis as well as multidimensional LC followed by MS for subcellular organelle isolation, protein fractionation, and separation. Methods including two-dimensional difference gel electrophoresis (2D-DIGE), isotope-coded affinity tag (ICAT), and isobaric tag for relative and absolute quantitation (iTRAQ) were compared for quantifying relative protein profiles among samples, and an overview evaluation of these techniques has been published (10).

In our study, to establish the biological difference between atenolol-incubated A7r5 cells and control A7r5 cells, the protein profile in A7r5 cells incubated with individual enantiomers of atenolol was analyzed by 2D LC-MS/MS (11); the procedures were shown in Fig. 1. More than 200 proteins were identified from each of the three independent experiments conducted by the following criteria: unused protein score was more than 2 (99 % confidence) per experiment, and one or more peptide hits were found per protein at >95 % confidence per peptide.

2 Materials

Prepare all solutions using ultrapure water (prepared by purifying deionized water to attain a sensitivity of 18 MΩ cm at 25 °C) and analytical grade reagents. Prepare and store all reagents at room temperature (unless indicated otherwise). Diligently follow all waste disposal regulations when disposing waste materials.

1. 20 μM Atenolol.
2. Dulbecco's modified Eagle's medium was supplemented with 10 % fetal bovine serum, 100 U/ml penicillin, and 100 U/ml streptomycin.
3. Cell lysis buffer: 150 μl of 8 M urea, 4 % (w/v) CHAPS, and 0.05 % SDS (w/v).
4. BSA.
5. Acetone was pre-cooled at −20 °C.
6. 10, 20, 30, 40, 50, 60, 80, 100, 300, and 500 mM of KCl solution.
7. 0.1 % formic acid.
8. 95 % acetonitrile + 0.1 % formic acid.
9. 5 % acetonitrile + 0.1 % formic acid.

3 Methods

3.1 Cell Culture

Vascular smooth muscle cells, A7r5 cells obtained from the American Type Culture Collection, were cultured in Dulbecco's modified Eagle's medium. Cells were maintained at 37 °C in an atmosphere of 5 % CO_2. All culture media and media supplements were purchased from Invitrogen. After reaching 80 % confluence, the cells were incubated with the *S*-enantiomer and the *R*-enantiomer of atenolol at a concentration of 20 μM, respectively, for 24 h in the absence of serum (*see* **Note 1**).

3.2 Cell Lysis, Protein Digestion, and Labeling with iTRAQ Reagents

1. Cells of 10 cm petri dish were harvested and lysed in 150 μl of 8 M urea, 4 % (w/v) CHAPS, and 0.05 % SDS (w/v) on ice for 20 min with regular vortexing.
2. Protein was centrifuged at 15,000 × *g* for 60 min at 4 °C, supernatant was removed, and protein was quantified using the 2D Quant kit (GE Healthcare). A standard curve was made using BSA as a control (*see* **Note 2**).
3. A total of 100 μg of each sample was precipitated by the addition of four times the sample volume of cold (−20 °C) acetone to the tube for 2 h and centrifuged at 12,000 × *g* for 1 h at 4 °C, and then the supernatant was decanted.

4. The protein pellets were then dissolved in the solution buffer and denatured, and cysteines were blocked as described in the iTRAQ protocol (Applied Biosystems). To each sample tubes, add 20 μl Dissolution Buffer and 1 μl Denaturant, vortex and mix with samples.
5. To each sample tubes, add 2 μl Reducing Reagent, vortex and mix with samples, then incubate the tubes at 60 °C for 1 h.
6. To each tube, add 1 μl Cysteine Blocking Reagent, vortex and mix with samples, incubate the tubes at room temperature for 10 min (*see* **Note 3**).
7. Each sample is digested with 20 μl of 0.25 μg/μl sequencing grade modified trypsin (Promega) solution at 37 °C overnight (*see* **Note 4**).
8. The iTRAQ tags added 70 μl ethanol 1 h ago was used to label digested protein samples as follows: control A7r5, iTRAQ 114; A7r5 incubated with the *S*-enantiomer of atenolol, iTRAQ 115; and A7r5 incubated with the *R*-enantiomer of atenolol, iTRAQ 117.
9. The labeled samples were then pooled in a fresh tube before analysis.

3.3 Online 2D Nano-LC-MS/MS Analysis

The analysis was performed on an Agilent 1200 nanoflow LC system (Agilent Technologies) interfaced with a QSTAR XL mass spectrometer (Applied Biosystems/MDS Sciex).

1. 20 μl iTRAQ-labeled samples were added in an HPLC vial; put vial on loading tray in HPLC machine (*see* **Note 5**).
2. After samples were injected in HPLC, in the first dimension, 3 μl of the combined peptide mixture was loaded onto the PolySulfoethyl A SCX column (0.32 × 50 mm, 5 μm) and was eluted stepwise by injecting salt plugs of ten different molar concentrations of 10, 20, 30, 40, 50, 60, 80, 100, 300, and 500 mM of KCl solution (*see* **Note 6**).
3. In the second dimension, while the 10-port valve being in position 1 (Fig. 2a), the sequentially eluted peptides from SCX column were trapped onto the ZORBAX 300SB-C18 enrichment column 1 (0.3 × 5 mm, 5 μm) and washed isocratically with buffer A (5 % acetonitrile, 0.1 % formic acid) at 0.005 ml/min for 100 min to remove any excess reagent. Meanwhile, the peptides bound on the ZORBAX 300SB-C18 enrichment column 2 during the previous run were eluted with buffer B (0.1 % formic acid) and buffer C (a nanoflow gradient of 5–80 % acetonitrile + 0.1 % formic acid) at a flow rate of 500 nl/min. Further separation was achieved onto the analytical Zorbax 300SB C-18 reversed-phase column (75 μm × 50 mm, 3.5 μm). In the next run, while the 10-port valve being switched to position 2 (Fig. 2b), the column 1 was switched into the solvent path of the nanopump, and the column 2 was used to trap the newly eluted peptides from SCX.

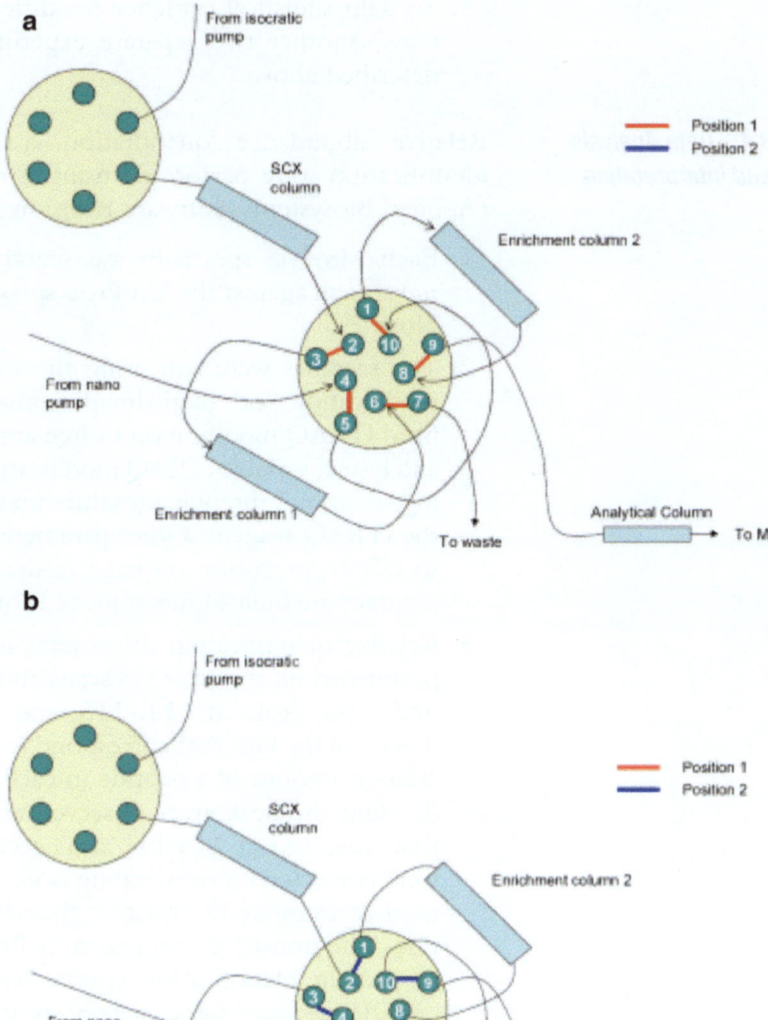

Fig. 2 Flow diagram for online nano-2D LC. (**a**) The 10-port valve was in position1: bypassing of SCX column, washing of enrichment column 1, and analysis of peptides from enrichment column 2. (**b**) The 10-port valve was in position 2: bypassing of SCX column, washing of enrichment column 2, and analysis of peptides from enrichment column 1

4. Altogether 11 runs were performed to finish one experiment. For MS/MS analysis, survey scans were acquired from m/z 300–1,500 with up to two precursors selected for MS/MS from m/z 100–2,000 using dynamic exclusion, and the rolling collision energy was used to promote fragmentation.

5. To gain statistical evidence for differential expression of proteins, another two separate experiments were performed as described above.

3.4 Data Analysis and Interpretation

Relative abundance quantitation and peptide and protein identification were performed using ProteinPilot™ Software 2.0 (Applied Biosystems, Software Revision 50861).

1. Each MS/MS spectrum was searched for species of *Rattus norvegicus* against the UniProt_sprot_20070123 database (*see* **Note 7**).
2. The searches were run using the following parameters: fixed modification of methylmethanethiosulfate-labeled cysteine, fixed iTRAQ modification of free amine in the amino terminus and lysine, variable iTRAQ modifications of tyrosine, and allowing serine and threonine residues undergoing side reaction with the iTRAQ reagent. Other parameters such as tryptic cleavage specificity, precursor ion mass accuracy, and fragment ion mass accuracy are built-in functions of ProteinPilot software.
3. Relative quantification of proteins in the case of iTRAQ was performed on the MS/MS scans and was the ratio of the areas under the peaks at 114, 115, and 117 Da, which were the masses of the tags that correspond to the iTRAQ reagents. The relative amount of a peptide in each sample was calculated by dividing the peak areas observed at 115.1 and 117.1 *m/z* by that observed at 114.1 *m/z*. The calculated peak area ratios were corrected for overlapping isotopic contributions and were used to estimate the relative abundances of a particular peptide. The unused protein score is ProteinPilot's measurement of protein identification confidence taking into account all peptide evidence for a protein, excluding any evidence that is better explained by a higher ranking protein. Sequence coverage was calculated by dividing the number of amino acids observed by the protein amino acid length.
4. The following criteria were required to consider a protein for further statistical analysis: two or more high confidence (>95 %) unique peptides had to be identified and the -fold difference had to be greater than 1.2. The candidate proteins were carefully examined in the Protein ID of the ProteinPilot software. The peptides without any modification of free amine in the amino terminus or without iTRAQ modification of free amine in the lysine were excluded from calculation of the protein ratios. To account for small differences in protein loading, all protein ratios were normalized using the overall ratios for all proteins in the sample as recommended by Applied Biosystems.

A total of 361 unique proteins were identified in the three independent experiments. A total of 19 unique proteins which were differentially expressed between the Atenolol-incubated and normal samples were selected for further analysis.

4 Notes

1. 10 cm Petri dish plates were usually used to culture cells. However, 96 well plates may also be used in cell MTT assay.
2. For protein concentration measurement, 2D Quant method can be substituted by Bradford method.
3. Due to the presence of highly volatile mercaptoethanol in Cysteine Blocking Reagent, the blocking step should be carried out in a chemical hood.
4. The tube should be packed by parafilm tightly to prevent solution evaporation to prevent the samples from drying after overnight incubation.
5. The sample should be centrifuged at >12,000 × *g* or go through 3k filter (Millipore) to remove precipitates, which may clog column and damage the HPLC machine. Once the aliquot of samples was loaded on HPLC column, the remaining part of the samples should be kept at −25 °C till further usage.
6. HPLC setting and operation should be handled by trained researchers.
7. The species selected in analysis software should be consistent with the source of samples. Database can be updated online to acquire the newest version.

References

1. Wilkins MR, Sanchez JC, Gooley AA, Appel RD, Humphery-Smith I, Hochstrasser DF, Williams KL (1996) Progress with proteome projects: why all proteins expressed by a genome should be identified and how to do it. Biotechnol Genet Eng Rev 13:19–50
2. Aebersold R, Mann M (2003) Mass spectrometry-based proteomics. Nature 422:198–207
3. Sui J, Tan TL, Zhang J, Ching CB, Chen WN (2007) iTRAQ-coupled 2D LC-MS/MS analysis on protein profile in vascular smooth muscle cells incubated with S- and R-enantiomers of propranolol: possible role of metabolic enzymes involved in cellular anabolism and antioxidant activity. J Proteome Res 6:1643–1651
4. Zhang HT, Gorn M, Smith K, Graham AP, Lau KK, Bicknell R (1999) Transcriptional profiling of human microvascular endothelial cells in the proliferative and quiescent state using cDNA arrays. Angiogenesis 3:211–219
5. Zhang J, Sui J, Ching CB, Chen WN (2008) Protein profile in neuroblastoma cells incubated with S- and R-enantiomers of ibuprofen by iTRAQ-coupled 2-D LC-MS/MS analysis: possible action of induced proteins on Alzheimer's disease. Proteomics 8:1595–1607
6. Qian WJ, Jacobs JM, Liu T, Camp DG 2nd, Smith RD (2006) Advances and challenges in liquid chromatography-mass spectrometry-based proteomics profiling for clinical applications. Mol Cell Proteomics 5:1727–1744
7. Ducret A, Van Oostveen I, Eng JK, Yates JR 3rd, Aebersold R (1998) High throughput protein characterization by automated reverse-phase chromatography/electrospray tandem mass spectrometry. Protein Sci 7:706–719
8. Takahashi N, Kaji H, Yanagida M, Hayano T, Isobe T (2003) Proteomics: advanced technology for the analysis of cellular function. J Nutr 133:2090S–2096S

9. Shen Z, Wang M, Briggs SP (2006) Use of high-throughput LC–MS/MS proteomics technologies in drug discovery. Drug Discov Today Tech 3:301–306
10. Tannu NS, Hemby SE (2006) Methods for proteomics in neuroscience. Prog Brain Res 158:41–82
11. Sui J, Zhang J, Tan TL, Ching CB, Chen WN (2008) Comparative proteomics analysis of vascular smooth muscle cells incubated with S- and R-enantiomers of atenolol using iTRAQ-coupled two-dimensional LC-MS/MS. Mol Cell Proteomics 7:1007–1018

Chapter 5

Phosphoproteomic Analysis of Aortic Endothelial Cells Activated by Oxidized Phospholipids

Alejandro Zimman, Judith A. Berliner, and Thomas G. Graeber

Abstract

Comprehensive identification of quantitative changes in protein phosphorylation using mass spectrometry is becoming a common tool in cell signaling studies. To date, most of these kinase network studies are conducted in stable cancer cell lines, yeasts, or other models that are not representative of cardiovascular disease. We describe methods based on phosphopeptide enrichment after tryptic digestion of cell lysates to study changes in protein phosphorylation of endothelial cells. We used this approach to study the activation of aortic endothelial cells by oxidized phospholipids, compounds important in atherosclerosis and other inflammatory diseases.

Key words Atherosclerosis, Endothelial cell, Oxidized phospholipid, Ox-PAPC, Kinase signaling, Phosphoproteomic profiling, Phosphopeptide enrichment, Phosphatase inhibitor

1 Introduction

Atherosclerosis is a major contributor to heart disease and stroke characterized by lipid accumulation in the vessel wall, monocyte entry, and in the final stages plaque rupture leading to thrombosis. We and others have shown that oxidized phospholipids play an important role in atherosclerosis by activating endothelial-monocyte interaction and other pro-atherogenic processes (1, 2). In a study of human aortic endothelial cells from 150 human donors, over 1,000 genes were shown to be regulated by oxidized products of the phospholipid 1-palmitoyl-2-arachidonyl-*sn*-3-glycero-phosphorylcholine (Ox-PAPC) (3). Oxidized phospholipids are found in atherosclerotic lesions (4), associated with LDL particles, and in apoptotic cells (1, 5). The levels of oxidized phospholipids in plasma are prognostic for atherosclerosis (6). We and others have employed different approaches to understand the mechanisms involved in endothelial cell activation by these lipids in the early stages of atherosclerosis (1–3). To further our understanding on the molecular mechanism(s) responsible for the activation of

Fernando Vivanco (ed.), *Vascular Proteomics: Methods and Protocols*, Methods in Molecular Biology, vol. 1000, DOI 10.1007/978-1-62703-405-0_5, © Springer Science+Business Media New York 2013

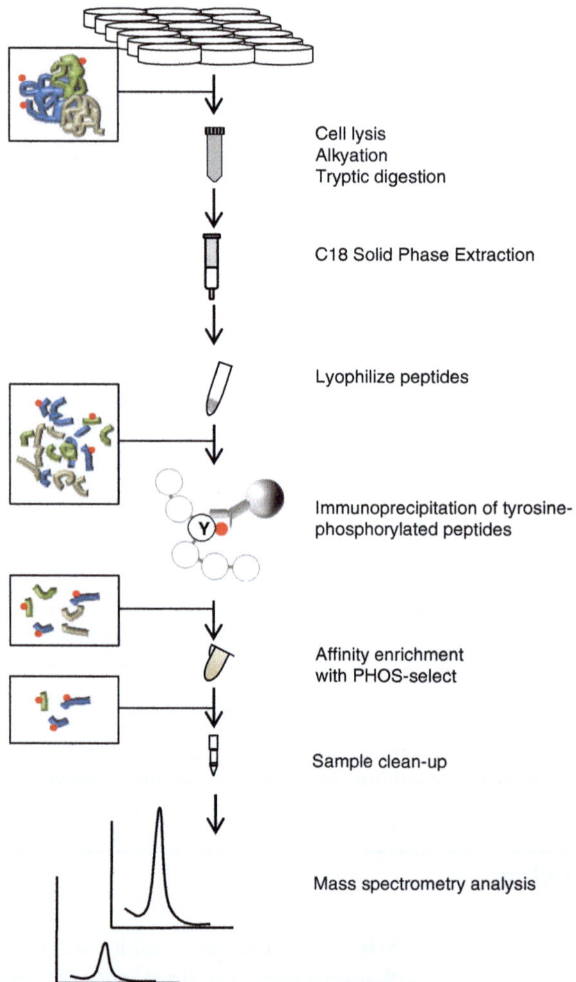

Fig. 1 Experimental scheme of immunoprecipitation of tyrosine-phosphorylated peptides and enrichment with iron metal affinity media

endothelial cells by oxidized phospholipids, we implemented phosphoproteomic methods for the analysis of aortic endothelial cells (7).

The detailed methods described in this chapter are based on the enrichment of phosphopeptides obtained after tryptic digestion of protein lysates to identify changes in phosphorylation sites by Ox-PAPC. Two different approaches are presented, which either target phosphotyrosine peptides or phospho-serine/threonine/peptides. For the phosphotyrosine-targeted approach, we performed immunoprecipitation with a pan-specific anti-phosphotyrosine antibody (8–12) followed by phosphopeptide enrichment using PHOS-Select iron metal affinity gel (Fig. 1). For the phospho-serine/threonine-targeted approach, we fractionated the peptides by strong cation exchange (SCX) before enrichment of phosphopeptides with TiO_2 (10, 11)

(Fig. 2). Because of the lower abundance of tyrosine-phosphorylated proteins compared to serine/threonine, the SCX/TiO_2 approach will typically yield only a small percentage of tyrosine-phosphorylated peptides. In our application, we developed a strategy to use SCX in solid phase extraction (SPE) cartridge format instead of the more laborious and equipment-dependent SCX-HPLC separation. A recent publication demonstrated SCX performed in SPE can detect over 8,000 phosphorylation sites in yeast (13).

Results obtained by phosphoproteomic analysis of aortic endothelial cells activated by oxidized phospholipids yielded new important information. Endothelial cells are exposed to many activators and methods described here can be applied to other ligands, stimulants, and biological processes to study signaling pathways relevant to the endothelium.

2 Materials

Water used to dissolve chemicals refers to ultrapure water (18.2 MΩ cm at 25 °C).

2.1 Components for Treatment of Bovine Aortic Endothelial Cells with Oxidized Phospholipids and Lysis

1. Bovine aortic endothelial cells (VEC Technologies, Rensselaer, NY).
2. Media culture: MCDB131 (Invitrogen, Carlsbad, CA) supplemented with 15 % FBS (HyClone, Logan, UT), 90 μg/mL heparin (Sigma, St. Louis, MO), and 20 μg/mL endothelial cell growth supplement (Fisher Scientific, Pittsburgh, PA).
3. PAPC (1-palmitoyl-2-arachidonyl-*sn*-3-glycero-phosphorylcholine; Avanti Polar Lipids, Alabaster, AL).
4. Urea lysis buffer: 8 M Urea, 50 mM Tris–HCl (pH 7.5), 1 mM Na_3VO_4, 100 nM Okadaic acid (*see* **Note 1**).
5. Sonicator with dismembrator probe.
6. Centrifuge with swing bucket rotor that holds 15 and 50 mL plastic tubes and speeds up to 4,000 × *g*.
7. Syringe filters units (0.45 μm and 0.22 μm, 33 mm diameter).

2.2 Components for Alkylation, Digestion, and Reverse Phase Extraction

1. Sodium phosphate: 1 M sodium phosphate (pH 7.5). Dissolve 4.4 g sodium phosphate monobasic-monohydrate ($NaH_2PO_4 \cdot H_2O$; MW 137.99) and 45.1 g sodium phosphate dibasic-heptahydrate ($Na_2HPO_4 \cdot 7H_2O$; MW 268.07) in 200 mL water. Warm up to 37 °C to dissolve completely.
2. Reducing agent: 1 M DL-Dithiothreitol (DTT) in water.
3. Alkylating agent: 0.5 M Iodoacetamide in water. Prepare fresh and avoid repeated freeze/thaw.
4. Slide-A-Lyzer Dialysis Cassette Kit, 3.5 K MWCO (Pierce, Rockford, IL).

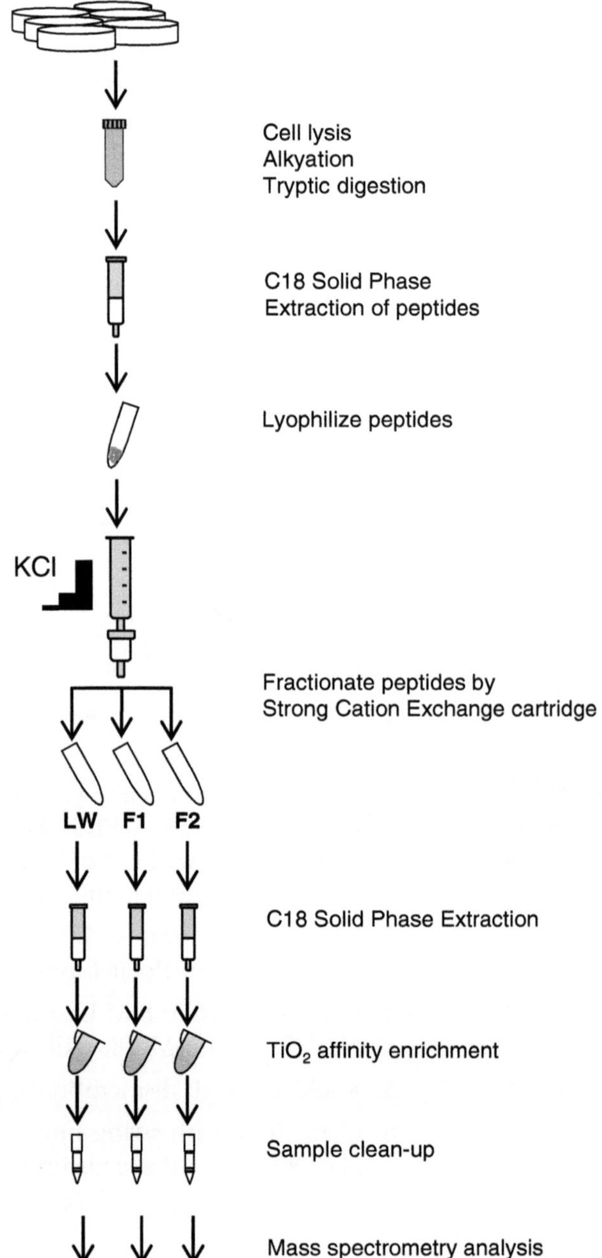

Fig. 2 Experimental scheme of affinity enrichment of serine/threonine-phosphorylated peptides

5. Dialysis buffer: 2 M Urea, 50 mM Tris base, glacial acetic acid to pH 8.0.
6. TPCK-treated trypsin (catalog number LS003740; Worthington, Lakewood, NJ): 5 μg/μL trypsin dissolved in buffer provided in the Sequencing Grade Modified Trypsin.

7. Sequencing Grade Modified Trypsin (catalog number V5113, Promega, Madison, WI).
8. Amicon Ultra-15 Centrifugal Filter Unit with Ultracel-10 membrane (catalog number UFC901008, Millipore, Billerica, MA).
9. Trifluoroacetic Acid (TFA) 10×1 mL-Ampule (catalog number 28904, Pierce, Rockford, IL). For 5 % solution, place 19 mL of water in a glass bottle with a wide opening. Open TFA ampoule inside a chemical hood and add it to the bottle containing water (*see* **Note 2**).
10. pH test papers range 0.0–5.5 and 6–9.5.
11. Discovery® DSC-18 SPE tube (500 mg bead weight, 3 mL volume, catalog number 52603-U, Sigma, St. Louis, MO).
12. Vacuum manifold.
13. Lyophilizer.

2.3 Components for Immunoprecipitation of Tyrosine-Phosphorylated Peptides and Enrichment with Iron Metal Affinity Media

1. Resuspension buffer: 100 mM Tris–HCl (pH 8).
2. Binding buffer: 50 mM Tris–HCl (pH 7.4).
3. 1.7 and 0.6 mL microcentrifuge tubes, low retention.
4. 1 M Tris base in water (untitrated).
5. Anti-Phosphotyrosine, clone 4 G10®, agarose conjugate (catalog number 16–101, Millipore, Billerica, MA).
6. Pierce Spin Cups (catalog number 69700, Pierce, Rockford, IL).
7. Vacuum concentrator or SpeedVac (*see* **Note 3**).
8. Sample solution: 250 mM acetic acid, 30 % acetonitrile.
9. Elution solution: 1.6 % NH_3 in water.
10. PHOS-Select™ Iron Affinity Gel (catalog number P9740, Sigma, St. Louis, MO).
11. Ultrafree-MC Centrifugal Filter (catalog number UFC30GV25, Millipore, Billerica, MA).

2.4 Components for Affinity Enrichment of Serine/Threonine-Phosphorylated Peptides

1. Buffer A: 5 mM KH_2PO_4 (pH 2.65), 30 % acetonitrile, 5 mM KCl (*see* **Note 4**).
2. Buffer B: 5 mM KH_2PO_4 (pH 2.65), 30 % acetonitrile 17.5 mM KCl (*see* **Note 4**).
3. Buffer C: 5 mM KH_2PO_4 (pH 2.65), 30 % acetonitrile 70 mM KCl (*see* **Note 4**).
4. PolySulfoethyl A in SPE cartridges (catalog number SPESE1203, Poly LC, Columbia, MD) (*see* **Note 5**).
5. 0.2 M sodium phosphate buffer (pH 7.5).
6. Equilibration buffer: 45 % acetonitrile 0.1 % TFA, 150 mg/mL lactic acid (catalog number 69785, Fluka).
7. Rinsing buffer: 45 % acetonitrile 0.1 % TFA.

8. Elution solution: 3 % NH_3 in water.
9. Titanium dioxide (catalog number TT200TIO, Poly LC, Columbia, MD) (*see* **Note 6**).

2.5 Components for Sample Clean-Up for Mass Spectrometry Analysis

1. MonoTip C18, 200 μL volume (catalog number 1401096, GL Sciences, Torrance, CA) (*see* **Note 7**).
2. Activation solution: 100 % acetonitrile.
3. Wash solution: 0.1 % TFA in water.
4. Elution solution: 0.1 % TFA, 50 % acetonitrile in water.

2.6 Mass Spectrometry Analysis

1. LC-MS/MS using, for example, an Eksigent autosampler coupled with Nano2DLC pump (Eksigent, Dublin, CA) and LTQ-Orbitrap (Thermo Fisher Scientific, Waltham, MA).
2. Analytical column (10 cm×75 μm i.d.) packed with 5 μm Integrafit Proteopep2 300 Å C18 (New Objective, Woburn, MA).
3. HPLC solvents: buffer A (0.1 % formic acid in water) and buffer B (0.1 % formic acid in acetonitrile). All HPLC solvents are Ultima Gold quality (Fisher Scientific).

3 Methods

The method describes the phosphoproteomic analysis of primary aortic endothelial cells to identify changes in tyrosine, serine, and threonine phosphorylation between two conditions (media alone vs. Ox-PAPC). Focused analysis of tyrosine phosphorylation is separate from serine/threonine phosphorylation due to lower abundance of tyrosine-phosphorylated proteins and the availability of highly pan-specific and sensitive anti-phosphotyrosine antibodies. It is possible to identify tyrosine, serine, and threonine phosphorylation from the same sample through sequential enrichment steps (*see* **Note 14**) or to target only tyrosine phosphorylation or serine/threonine phosphorylation. These methods have been successfully applied to other cell types.

We co-treated endothelial cells with phosphatase inhibitors to facilitate phosphoproteomic analysis. This optional step sustains phosphorylation events and amplifies differences between Control and ligand, resulting in more leads for the signaling stimuli being studied. However, phosphatase inhibitors could also create artifacts in the signaling network. Thus, inhibitor-derived results should be validated in inhibitor-free follow-up studies.

3.1 Treatment of Primary Aortic Endothelial Cells with Oxidized Phospholipids

1. PAPC is oxidized to Ox-PAPC by exposure to air and stored in chloroform as previously described (4).
2. Grow bovine aortic endothelial cells in 150 mm dishes with culture media and treat with lipids at ~90 % confluence. For phosphotyrosine enrichment (Subheading 3.3), we used 18

dishes (approximately 30 mg of protein) per sample. For phospho-serine/threonine enrichment (Subheading 3.4), we used six dishes (approximately 10 mg of protein per sample) (*see* **Note 8**).

3. Remove the growth media and add 40 μg/mL of Ox-PAPC dissolved in MCDB131 media with 1 % FBS containing 1 mM vanadate (for tyrosine phosphorylation) or 100 nM okadaic acid (for serine/threonine phosphorylation) for 40 min. In the case of serine/threonine phosphorylation, cells were pretreated with 100 nM okadaic acid for 1 h. For Control experiments, incubate cells with the baseline media preparation containing phosphatase inhibitors (*see* **Note 9**).
4. Remove the media containing the lipid and wash with ice-cold PBS. Then add more ice-cold PBS, scrape the cells off the plate, and place in a 50 mL conical tube. In some instances, including phosphatase inhibitors in the PBS can help sustain phosphorylation events. The benefit of this option can be studied by immunoblotting (*see* **Note 10**).
5. Pellet cells at 2,000 × *g* for 5 min.
6. Resuspend the pellet with 20 mL (for phosphotyrosine enrichment, Subheading 3.3) or 10 mL (for phospho-serine/threonine, Subheading 3.4) urea lysis buffer and vortex vigorously.
7. Sonicate cell lysate until homogenous (no clumps). Sonication conditions will depend on the probe performance and cell characteristics. For reference, we used three rounds of sonication for 1 min each. Make sure the lysate remains ice-cold during sonication.
8. Spin the lysate at 3,500 × *g* for 15 min at room temperature.
9. Filter supernatant through 0.45 μm filter units followed by 0.22 μm. Filter unit may get clogged. Replace filter units as needed. Measure the sample volume accurately (one method is to pipette the sample in 10 mL plastic pipette to measure the volume).
10. Determine protein concentration using Bradford assay (*see* **Note 11**).
11. Aliquot 100 μL of sample for immunoblotting and freeze the remaining sample at −80 °C (*see* **Note 10**).

3.2 Alkylation, Digestion, and Reverse Phase Extraction

1. Thaw the samples and add sodium phosphate (pH 7.5) to a final concentration of 100 mM.
2. Add 1 M DTT to each sample to a final concentration of 5 mM. Incubate in water bath for 1 h at 37 °C.
3. Add 0.5 M iodoacetamide to each sample to a final concentration of 25 mM. Incubate in the dark for 45 min at room temperature.

4. Add 1 M DTT to each sample to a final concentration of 10 mM. Incubate for 30 min at room temperature.
5. Place the sample inside a Slide-A-Lyzer Dialysis cassette (3.5 K MWCO) and dialyze for 4 h in 2 L of dialysis buffer. Replace the buffer after 2 h (*see* **Note 12**).
6. Recover the sample from the cassette and add TPCK-treated trypsin at 1:100 trypsin/protein ratio. Check the pH is 7.5–8 with 1–2 μL sample in pH test papers and add 1 M Tris base, if necessary. Incubate for 2–4 h at 37 °C.
7. Add Sequencing Grade Modified Trypsin at 1:100 trypsin/protein ratio. Check again that the pH is 7.5–8. Incubate overnight at 37 °C mixing continuously (*see* **Note 13**).
8. Filter digested lysate through 0.22 μm filter.
9. Filter again by centrifugation with Amicon Ultra-15 Centrifugal Filter Unit until volume on top of filter is less than 0.25 mL. Collect the flow through.
10. Acidify the flow through containing tryptic peptides with 5 % TFA to pH 3.5–4.5. Check the pH with 1–2 μL sample using pH test papers.
11. Activate Discovery® DSC-18 SPE tube by connecting the cartridge to a vacuum manifold and washing with two columns of acetonitrile followed by two columns of 0.1 % TFA in water. Do not let the SPE tube to dry. Use a slow flow rate (i.e., one drop per second) to allow proper exchange of the sample with the C18 material.
12. Load the sample into the C18 SPE tube.
13. Wash with two column volumes of 0.1 % TFA in water.
14. Elute with 40 % acetonitrile, 0.1 % TFA. Collect 4 × 2 mL fractions into glass tubes. Label each tube to identify the sample and elution fraction. Discard the C18 SPE tube.
15. Cover the tubes and freeze the solution by placing them in powdered dry ice. After the solution is completely frozen, puncture the cover and place the tubes in the lyophilizer overnight.
16. Samples are ready for phosphopeptide enrichment described in Subheadings 3.3 or 3.4.

3.3 Immunoprecipitation of Tyrosine-Phosphorylated Peptides and Enrichment with Iron Metal Affinity Media

1. Dissolve lyophilized peptides in resuspension buffer, transfer to a low retention tube, and adjust pH to 7.4 with 1 M Tris base. Vortex vigorously to dissolve all material and add more buffer if cloudy.
2. Wash 4G10 antibody conjugated to agarose with ice-cold binding buffer twice. Washes are done by centrifugation at 960 × *g*, removal of the supernatant, addition of buffer, and

gentle vortex mixing. After the final wash, add buffer so that the concentration is back to 50 % slurry.

3. Add 250 μL (50 % slurry) pre-washed 4G10 beads to peptides obtained from tryptic digestion of 30 mg of protein from BAEC. Incubate overnight at 4 °C on the rotator. Other conditions or cell types may require optimization for the amount of antibody needed.
4. After immunoprecipitation, transfer 400 μL of sample to Pierce Spin Cups and centrifuge at 960 × *g* for 1 min. Repeat until all material is transferred to the cup (*see* **Note 14**).
5. Wash the beads with 450 μL binding buffer three times.
6. Wash the beads with 450 μL 25 mM NH_4HCO_3 twice.
7. Resuspend the beads with 200 μL 25 mM NH_4HCO_3 and transfer to a clean 1.7 mL microcentrifuge tube. Rinse twice the Spin Cup with 200 μL 25 mM NH_4HCO_3 and transfer each rinse to the same microcentrifuge tube.
8. Spin at 1,700 × *g* for 1 min and remove the supernatant completely by dipping a gel loading tip slightly below the bead surface.
9. Add 500 μL 0.1 % TFA to the 4G10 beads. Mix well and incubate for 15 min at 37 °C. Vortex at least every 5 min to keep the beads in suspension.
10. Transfer the entire suspension to an Ultrafree-MC Centrifugal Filter (*see* **Note 15**). Centrifuge the filter at 960 × *g* for 1 min. Add 100 μL 0.1 % TFA to the dried beads, mix gently, and centrifuge again.
11. Concentrate the sample by vacuum centrifugation until dry (*see* **Note 16**).
12. Dissolve dried phosphotyrosine peptides in 500 μL sample solution.
13. Wash PHOS-Select slurry with sample solution three times. Washes are done by centrifugation at 2,700 × *g* for 1 min, removal of the supernatant, addition of buffer, and gentle vortex. After the final wash, add buffer so that the concentration is back to 50 % slurry.
14. Add 60 μL PHOS-Select (50 % slurry) to the peptides obtained from tryptic digestion and phosphotyrosine immunoprecipitation of 30 mg of protein from BAEC. Incubate for 45 min at room temperature rotating.
15. Transfer sample to an Ultrafree-MC Centrifugal Filter (*see* **Note 15**). Centrifuge the filter at 2,700 × *g* for 1 min. Discard flow through.
16. Wash PHOS-Select beads with 250 μL sample solution twice. Use these washes to clean the tube used for the enrichment with PHOS-Select.

17. Wash PHOS-Select beads with 250 μL water. Replace the microcentrifuge tube to collect eluate.
18. Add 250 μL elution buffer to cover all PHOS-Select beads and incubate for 5 min at room temperature. Centrifuge at 2,700 × *g* for 30 s.
19. Add another 100 μL elution buffer to the PHOS-Select beads and centrifuge at 2,700 × *g* for 30 s.
20. Concentrate combined eluates by vacuum centrifugation until dry. Clean up samples for mass spectrometry as described in Subheading 3.5.

3.4 Affinity Enrichment of Serine/Threonine-Phosphorylated Peptides

1. Dissolve lyophilized peptides in 2 mL buffer A.
2. Connect PolySulfoethyl A SPE cartridge to a 5 mL syringe.
3. Activate PolySulfoethyl A with 4 mL acetonitrile followed by 4 mL buffer A.
4. Load sample and start collecting the eluate in glass tubes right away.
5. When almost all the sample has passed through the cartridge (do not let the column to dry), add 2 mL buffer A and pool with the first 2 mL collected. This will be fraction LW (Load & Wash).
6. Collect the next fraction (F1) with 4 mL of buffer B.
7. Collect the final fraction (F2) with 4 mL of buffer C.
8. Evaporate acetonitrile by vacuum centrifugation (i.e., volume in each tube is less than 2.8 mL if starting volume is 4 mL). Do not dry the sample.
9. Without adjusting the pH, remove salts with C18 SPE tube as indicated in Subheading 3.2, **steps 11–13**. Elute all fractions in 4 mL 50 % acetonitrile, 0.1 % TFA.
10. Add lactic acid to all fractions to a final concentration 150 mg/mL.
11. Place all TiO_2 material in a microcentrifuge tube and wash with 500 μL acetonitrile, followed by 0.2 M sodium phosphate (pH 7.5), and equilibration buffer twice. TiO_2 material is very dense and will not remain in suspension. For the washes, mix thoroughly by pipetting. Afterwards, split the TiO_2 material into microcentrifuge tubes and check that the volume of material is equal (*see* **Note 17**).
12. Add TiO_2 material to each fraction and mix continuously (keep beads in suspension) for 45 min at room temperature.
13. Centrifuge at slow speed and transfer the TiO_2 beads to a 1.7 microcentrifuge tube.
14. Wash TiO_2 beads with 300 μL equilibration buffer three times.

15. Wash TiO_2 beads with 300 μL rinsing buffer twice. While performing these washes, transfer the beads to an Ultrafree-MC Centrifugal Filter (*see* **Note 15**). Replace the microcentrifuge tube to collect eluate.
16. Add 100 μL elution solution and incubate for 5 min at room temperature. Centrifuge at 2,700 × *g* for 30 s. Repeat with another 100 μL elution solution.
17. Concentrate eluate by vacuum centrifugation until dry. Clean up samples for mass spectrometry as described in Subheading 3.5.

3.5 Sample Clean-Up for Mass Spectrometry Analysis

1. Dissolve dried phosphopeptides in 200 μL 0.1 % TFA. Make sure all material is dissolved. Add more buffer if necessary.
2. Activate MonoTip C18 (200 μL volume) with 180 μL acetonitrile three times followed by 180 μL 0.1 % TFA three times (*see* **Note 18**).
3. Pipette the sample containing phosphopeptides through the MonoTip C18 (25 times). Make sure the tip does not dry.
4. Wash the tip five times by pipetting 180 μL 0.1 % TFA and releasing the liquid into the waste for each wash. Make sure the tip does not dry.
5. Aliquot 200 μL 50 % acetonitrile, 0.1 % TFA into a 0.6 microcentrifuge tube (low retention). Elute peptides bound in the tip by pipetting the elution solution ten times. Make sure to release all the liquid back into the tube.
6. Dry the sample by vacuum centrifugation.

3.6 Mass Spectrometry and Phosphopeptide Identification

1. Dissolve each sample in 12 μL buffer A. It is critical to dissolve the sample in the same volume to quantitatively compare peptide phosphorylation between samples. Based on the method described in Subheading 3.3, 30 mg of protein used as starting material is enough for 6 replicate runs (2 μL per injection). The same number of runs can be done from 10 mg of protein used in the method described in Subheading 3.4 (*see* **Note 19**).
2. Run each sample in duplicate alternating all samples processed with the same method (*see* **Note 20**). Increasing the number of samples and replicate runs increases the chances to identify more unique phosphopeptides. Duplicate runs also allow the application of statistical methods to access the reproducibility and confidence levels of the quantitative differences.
3. Run biological replicates in the same batch of mass spectrometry runs whenever possible (*see* **Note 21**).
4. Analyze phosphorylated peptides by LC-MS/MS. We used an Eksigent autosampler coupled with NanoLC 2D pump (Eksigent, Dublin, CA) and LTQ-Orbitrap. Load samples

onto an analytical column (10 cm × 75 μm i.d.) packed with 5 μm Integrafit Proteopep2 300 Å C18. Elute peptides into the mass spectrometer using a HPLC gradient of 5–40 % buffer B in 45 min followed by a quick gradient of 40–90 % buffer B in 10 min, where buffer A contains 0.1 % formic acid in water and buffer B contains 0.1 % formic acid in acetonitrile. Mass spectra are collected in positive ion mode using the Orbitrap for parent mass determination and the LTQ for data-dependent MS/MS acquisition of the top five most abundant peptides.

5. Search MS/MS fragmentation spectra using SEQUEST against a bovine IPI protein database (downloaded from ftp.ebi.ac.uk). Search parameters include carboxyamidomethylation of cysteine as static modification. Dynamic modifications include phosphorylation on serine, threonine, tyrosine, and oxidation on methionine. Filter the results derived from database search using the following criteria: XCorr 1.0(+1), 1.5(+2), 2(+3), peptide probability score (BioWorks 3.2, Thermo Fisher Scientific) 0.001, and delta Cn 0.1. Use an additional filter of measured peptide mass tolerance, or mass accuracy (parts per million, ppm), in conjunction with both forward and reverse database searches (IPI bovine reversed using BioWorks) to keep the false discovery rate (FDR) below 2 % (*see* **Note 22**).
6. Apply the Ascore or other algorithm to more accurately localize the phosphate on the peptide (http://ascore.med.harvard.edu).

3.7 Chromatography Profile Alignment and Alignment-Based Peak Identification

1. Locate each phosphopeptide peak sequenced in some samples but not others in the remaining samples by aligning the chromatogram elution profiles using a dynamic time warping algorithm or equivalent (14–17) (*see* **Note 23**).
2. Use analysis software to extract peak heights or integrate peak area intensities (15, 17, 18).
3. Visually inspect each phosphopeptide peak across all samples using extracted ion current (XIC) chromatography plots to ensure proper peak localization for height determination or integration.
4. Calculate the ratio of phosphopeptide peak intensities, dividing the average peak intensity from Ox-PAPC-treated cells by the average peak intensity from Control cells. Add together the peak intensities of phosphopeptides eluting in more than one fraction. Alternatively, use integrated peak areas.
5. Apply statistical tests. For example, a *t*-test *p*-value can be calculated between the replicate runs of LC-MS analysis of Control and Ox-PAPC samples. If the same phosphopeptide peak is quantified in two biological replicates, then the lowest ratio and the maximum *t*-test *p*-value of the pair of biological replicates is

retained. If the ratios of two biological replicates have opposing trends, then assign a ratio of 1 with no *t*-test *p*-value. If a phosphopeptide peak is quantified in only one biological replicate, then use the single ratio and *t*-test *p*-value available.

6. Determine the chance of obtaining false-positive fold-change results by comparing normalized values from the same treatment between biological replicates (i.e., Control vs. Control and Ox-PAPC vs. Ox-PAPC). The intensity values for the same phosphopeptide are scaled to each other by setting the mean of each peptide across the samples for one experimental batch equal to the mean for that peptide in the second experimental batch. Establish cutoff thresholds for ratio and *t*-test *p*-value so that the false-positive rate in the final results is less than 1 %. These cutoff values are calculated independently for phosphopeptides detected and quantified in either one or two biological replicates. In our work, phosphopeptides with the same sequence but different charge states or methionine oxidation states were required to have all of the states above cutoff values to be considered as induced by Ox-PAPC.
7. Sequence-align bovine IPI sequences using Blast against the human IPI protein sequence database to conduct further data analysis using the human homologue, including the identification of the corresponding phosphorylation site.

4 Notes

1. An elaborate preparation of vanadate is suggested in many protocols (19). In our experience, the addition of vanadate to the media (100 mM stock solution prepared by dissolving the solid in PBS and filtered through 0.22 μm) or lysis buffer (solid added to a final concentration of 1 mM) is enough to prevent dephosphorylation in aortic endothelial cells. The high concentration of urea in the lysis buffer will denature proteins, including phosphatases, thus preventing enzymatic dephosphorylation.
2. Use of single-use ampoules facilitates the preparation of TFA solution and eliminates the risk of chemical spill.
3. An underperforming vacuum concentrator will significantly slow down sample preparation.
4. Adjust pH to 2.65 with 1 N HCl before adding acetonitrile. Precise buffer pH is very important because phosphopeptide fractionation in SCX is based on this pH.
5. Alternatively, it is possible to purchase PolySulfoethyl A bulk media (catalog number BMSE03, Poly LC) and construct the SPE cartridge.

6. The material is included in a tip so that the sample passes through. We obtained better results by taking the titanium dioxide (TiO_2) out of the tips, conditioning, and adding to the samples as done in an immunoprecipitation. Vigorous mixing is necessary to keep the dense TiO_2 material in suspension.
7. We prefer to use large capacity tips such as these because they make the sample clean-up process easier. The large binding capacity ensures that the sample will not saturate the C18 material and the large volumes reduce errors by the operator. If samples are analyzed in a core facility, it is recommended to contact the facility for further advice.
8. The large amount of protein required to detect phosphorylation events precluded us from using primary human aortic endothelial cells. Therefore, we decided to use bovine aortic endothelial cells available in large quantities and with similar biological responses to their human counterparts.
9. Several variables affect the ability to detect changes in protein phosphorylation. We found that Ox-PAPC is a low level but chronic activator of endothelial cells. This produces relatively small changes in phosphorylation compared to the sharp increase induced by ligands such as growth factors. We co-treated oxidized lipids with phosphatase inhibitors (vanadate for tyrosine phosphorylation and okadaic acid for serine/threonine phosphorylation). This approach sustains the phosphorylation events, increases the levels of phosphorylation, and amplifies differences between Control and ligand. While we and others have found phosphatase inhibitors to be useful experimental tools, there are caveats of potentially altering the signaling network. Thus, inhibitor-derived results should be validated in inhibitor-free follow-up studies.
10. It is helpful to check that the treatment causes differences in signaling before proceeding with the proteomic analysis. We did immunoblotting with anti-phosphotyrosine antibodies to verify changes in tyrosine phosphorylation levels. We used PKC or PKA substrate anti-phospho motif antibodies to verify changes in serine and threonine phosphorylation levels. Antibodies that recognize particular phosphorylation sites known to be induced by the ligand may also be used. For example, Ox-PAPC increases the phosphorylation of Erk 1/2. However, we note that mass spectrometry phosphopeptide analysis can detect phosphorylation differences even in samples that appear identical in immunoblot analysis using pan-specific anti-phospho antibodies (8).
11. Accurate protein concentration determination is critical for comparison between samples in this method. Proper dilution and replicates are essential. This major source of error is rarely mentioned in proteomic protocols and also applies to label-based

quantitation experiments (e.g., SILAC, stable isotope labeling with amino acids in cell culture (20)).

12. Alternatively, sample can be diluted with 50 mM Tris (pH 8.0) to 2 M urea.
13. 1 mM $CaCl_2$ can be added to increase enzymatic activity. In addition, the efficiency of enzymatic digestion can be checked by SDS-PAGE and Coomassie staining of the sample before and after digestion.
14. It is possible to use the flow through from this step for the enrichment of phospho-serine/threonine peptides. To do this, acidify the flow through with TFA to pH 3.5–4.5 and clean up the sample with C18 SPE tube as described in Subheading 3.2, **steps 10–16**. The lyophilized material can now be used as indicated in Subheading 3.4.
15. Cut the hinge connecting the lid to the tube and transfer the lid and centrifugal filter into a 1.7 mL low retention microcentrifuge tube.
16. At this point, the sample can be analyzed by mass spectrometry. In our experience, the additional enrichment with PHOS-Select described in Subheading 3.3, **steps 12–20**, significantly reduces the amount of unphosphorylated peptides and increases the number of unique phosphopeptides detected by mass spectrometry.
17. Alternatively, aliquots of the TiO_2 material can be transferred to microcentrifuge tubes, conditioned individually, and applied to individual samples to facilitate equal splitting of the material.
18. We found the high binding capacity and large working volumes for these tips are very convenient. Nevertheless, it is recommended to contact the mass spectrometry facility prior to sample preparation as each facility will have their set of requirements for sample clean-up.
19. We normally perform two replicate runs and the leftover sample is saved in case of problems with LC-MS/MS analysis.
20. Samples processed with different sub-proteome enrichment methods will have different chromatography elution profiles and may complicate manual or automated chromatographic alignment.
21. Regardless of the proteomic method employed, biological replicates and statistical analysis are important to ensure that the phosphorylation events and quantitative changes are not artifacts or noise. Running biological replicates together will increase the number of unique phosphopeptides identified and will facilitate chromatographic alignment, quantitation, and measures of statistical confidence.

22. The FDR can be calculated as 2(nrev/(nrev + nfor)) where nfor is the number of unique peptides identified from the forward database and *n*rev is the number of unique peptide identified from a reversed database. Recent mass spectrometry sequencing algorithms will automatically calculate FDRs (e.g., Thermo Scientific Proteome Discoverer Software).
23. The quantitation of a large data set obtained by mass spectrometry remains a challenge and, in our experience, the major bottleneck for analysis. We employed computer algorithm that require specialized personnel. Software is available for non-isotopic and isotope-label methods but software issues may occur. If quantitation capabilities are limited, it is highly recommended to target tyrosine phosphorylation first. The abundance of tyrosine phosphorylation is considerably lower than for serine and threonine. In addition, many tyrosine phosphorylation sites identified by mass spectrometry are highly valuable for research purposes (i.e., receptor tyrosine kinases and other key kinases well described in the literature). Manual quantitation of 2–4 samples is feasible and could provide exciting new leads to the laboratory. Research laboratory-developed and commercial software have been recently released that make this process more user friendly, for example, via graphical user interfaces (GUIs) (15, 17, 18).

Acknowledgments

This work was supported by NIH grants HL030568 (to J.A.B.) and HG002807 (to T.G.G.). A. Z. received a postdoctoral fellowship from the American Heart Association (Western States).

References

1. Berliner JA, Leitinger N, Tsimikas S (2009) The role of oxidized phospholipids in atherosclerosis. J Lipid Res 50(Suppl):S207–S212
2. Bochkov VN, Oskolkova OV, Birukov KG, Levonen AL, Binder CJ, Stockl J (2010) Generation and biological activities of oxidized phospholipids. Antioxid Redox Signal 12: 1009–1059
3. Romanoski CE, Che N, Yin F, Mai N, Pouldar D, Civelek M, Pan C, Lee S, Vakili L, Yang WP, Kayne P, Mungrue IN, Araujo JA, Berliner JA, Lusis AJ (2011) Network for activation of human endothelial cells by oxidized phospholipids: a critical role of heme oxygenase 1. Circ Res 109:e27–e41
4. Watson AD, Leitinger N, Navab M, Faull KF, Horkko S, Witztum JL, Palinski W, Schwenke D, Salomon RG, Sha W, Subbanagounder G, Fogelman AM, Berliner JA (1997) Structural identification by mass spectrometry of oxidized phospholipids in minimally oxidized low density lipoprotein that induce monocyte/endothelial interactions and evidence for their presence in vivo. J Biol Chem 272: 13597–13607
5. Chang MK, Binder CJ, Miller YI, Subbanagounder G, Silverman GJ, Berliner JA, Witztum JL (2004) Apoptotic cells with oxidation-specific epitopes are immunogenic and proinflammatory. J Exp Med 200: 1359–1370
6. Tsimikas S, Brilakis ES, Miller ER, McConnell JP, Lennon RJ, Kornman KS, Witztum JL, Berger PB (2005) Oxidized phospholipids, Lp(a) lipoprotein, and coronary artery disease. N Engl J Med 353:46–57

7. Zimman A, Chen SS, Komisopoulou E, Titz B, Martinez-Pinna R, Kafi A, Berliner JA, Graeber TG (2010) Activation of aortic endothelial cells by oxidized phospholipids: a phosphoproteomic analysis. J Proteome Res 9:2812–2824
8. Skaggs BJ, Gorre ME, Ryvkin A, Burgess MR, Xie Y, Han Y, Komisopoulou E, Brown LM, Loo JA, Landaw EM, Sawyers CL, Graeber TG (2006) Phosphorylation of the ATP-binding loop directs oncogenicity of drug-resistant BCR-ABL mutants. Proc Natl Acad Sci USA 103:19466–19471
9. Rush J, Moritz A, Lee KA, Guo A, Goss VL, Spek EJ, Zhang H, Zha XM, Polakiewicz RD, Comb MJ (2005) Immunoaffinity profiling of tyrosine phosphorylation in cancer cells. Nat Biotechnol 23:94–101
10. Olsen JV, Blagoev B, Gnad F, Macek B, Kumar C, Mortensen P, Mann M (2006) Global, in vivo, and site-specific phosphorylation dynamics in signaling networks. Cell 127:635–648
11. Villen J, Beausoleil SA, Gerber SA, Gygi SP (2007) Large-scale phosphorylation analysis of mouse liver. Proc Natl Acad Sci USA 104:1488–1493
12. Rubbi L, Titz B, Brown L, Galvan E, Komisopoulou E, Chen SS, Low T, Tahmasian M, Skaggs B, Muschen M, Pellegrini M, Graeber TG (2011) Global phosphoproteomics reveals crosstalk between Bcr-Abl and negative feedback mechanisms controlling Src signaling. Sci Signal 4:ra18
13. Dephoure N, Gygi SP (2011) A solid phase extraction-based platform for rapid phosphoproteomic analysis. Methods 54:379–386
14. Prakash A, Mallick P, Whiteaker J, Zhang H, Paulovich A, Flory M, Lee H, Aebersold R, Schwikowski B (2006) Signal maps for mass spectrometry-based comparative proteomics. Mol Cell Proteomics 5:423–432
15. Park SK, Venable JD, Xu T, Yates JR 3rd (2008) A quantitative analysis software tool for mass spectrometry-based proteomics. Nat Methods 5:319–322, http://fields.scripps.edu/census
16. Vandenbogaert M, Li-Thiao-Te S, Kaltenbach HM, Zhang R, Aittokallio T, Schwikowski B (2008) Alignment of LC-MS images, with applications to biomarker discovery and protein identification. Proteomics 8:650–672
17. Cox J, Mann M (2008) MaxQuant enables high peptide identification rates, individualized p.p.b.-range mass accuracies and proteome-wide protein quantification. Nat Biotechnol 26:1367–1372, http://maxquant.org
18. Sadygov RG, Maroto FM, Huhmer AF (2006) ChromAlign: a two-step algorithmic procedure for time alignment of three-dimensional LC-MS chromatographic surfaces. Anal Chem 78:8207–8217, http://www.vastsci.com/resources/index.html
19. Huyer G, Liu S, Kelly J, Moffat J, Payette P, Kennedy B, Tsaprailis G, Gresser MJ, Ramachandran C (1997) Mechanism of inhibition of protein-tyrosine phosphatases by vanadate and pervanadate. J Biol Chem 272:843–851
20. Ong SE, Mann M (2006) A practical recipe for stable isotope labeling by amino acids in cell culture (SILAC). Nat Protoc 1:2650–2660

Chapter 6

Characterization of Membrane and Cytosolic Proteins of Erythrocytes

Gloria Alvarez-Llamas, Fernando de la Cuesta, Maria G. Barderas, Irene Zubiri, Maria Posada-Ayala, and Fernando Vivanco

Abstract

With the aim of studying a wide cohort of erythrocyte samples in a clinical setting, this chapter details a novel approach that allows the analysis of both human cytosolic and membrane sub-proteomes. Despite their simple structure, the high content of hemoglobin present in the red blood cells (RBCs) makes their proteome analysis enormously difficult. Careful investigation of different strategies for isolation of the membrane and cytosolic fractions from erythrocytes and their influence on proteome profiling by 2-DE was carried out, paying particular attention to hemoglobin removal. As result, a simple, quick, and satisfactory approach for hemoglobin depletion of erythrocyte cells based on HemogloBind™ reagent is shown here to satisfactorily analyze the cytosolic sub-proteome by 2-DE without major interference. For membrane proteome, a novel combined strategy based on hypotonic lysis isolation and further purification on minicolumns is described, allowing detection of high-molecular-weight proteins (i.e., spectrin, ankyrin) and well-resolved 2-DE patterns. The analysis of the membrane fraction by nano-LC coupled to an LTQ-Orbitrap mass spectrometer results in the identification of a total of 188 unique proteins.

Key words Erythrocytes, HemogloBind, Membrane proteins, Cytosolic proteins

1 Introduction

Membrane proteins actively participate in biological processes, such as surface recognition, cytoskeleton contact, signaling, enzymatic activity, or transporting, constituting one of the main targets for drugs in therapy treatments. In particular, they confer the red blood cells (RBCs) the adequate flexibility to fit into very small capillaries when supplying oxygen throughout the body. Cytosolic proteins may also be involved in a certain pathology contributing to the development of abnormalities in the diseased erythrocytes. Very recently, Goodman et al. (1) reviewed the up-to-now contributions to the knowledge of the erythrocyte proteome. A total of 751 proteins (membrane and cytosolic) have been identified by various proteomic approaches (2–4). In erythrocytes, the cell

Fernando Vivanco (ed.), *Vascular Proteomics: Methods and Protocols*, Methods in Molecular Biology, vol. 1000, DOI 10.1007/978-1-62703-405-0_6,

structure is very simple, lacking of nucleus and other organelles such as mitochondria, but the high content of hemoglobin may cause severe interference and obscure the detection of minor proteins. In particular, highly hydrophobic proteins, as those present in the cell membrane, are not the best candidate of choice in 2-DE due to their relative low abundance and high hydrophobicity (5). However, despite of its drawbacks, this technique is a good option in clinical proteomics in terms of ability to separate complex mixtures, display posttranslational modifications, and change after phosphorylation, among others. Analysis of the erythrocyte membrane sub-proteome has been described (6–10). However, little attention is usually paid to the simultaneous analysis of the cytosolic fraction of erythrocytes by 2-DE, probably due to the high content of hemoglobin present which enormously hampers its direct analysis. Here, we show a new strategy for 2-DE analysis of RBC membrane proteins, based on a commercial purification kit with variations. It results in a highly resolved spot pattern, and its application can be extended to other membrane proteomes analysis by 2-DE. Besides, a novel approach for hemoglobin depletion is provided, allowing the proteomic analysis of the cytosolic fraction in a quick and easy way.

2 Materials

2.1 Sample Collection and Fractions Isolation

1. EDTA tubes, plastic pipettes, and centrifugation tubes.
2. Stock buffer: 5 mM Na_2HPO_4, pH8.
3. Wash buffer: 0.9 % NaCl in stock buffer.
4. Lysis buffer: 1 mM EDTA, 1 mM PMSF (Sigma) in stock buffer freshly prepared.

2.2 Membrane and Cytosolic Fractions Pretreatment

1. HemogloBind™ reagent (Biotech Support Group LLC.).
2. Protein Desalting Spin Columns (Pierce).
3. 2-DE sample buffer: 7 M urea, 2 M thiourea, 4 % CHAPS.
4. 2-DE lysis buffer: 7 M urea, 2 M thiourea, 4 % CHAPS and 30 mM Tris.
5. Mem-PER® solution: mixture (2:1) of Mem-PER® Membrane Protein Solubilization Reagent (Pierce) and Mem-PER® buffer (Pierce); keep on ice all the time.
6. 2-D Sample Prep for Membrane Proteins kit (Pierce).

2.3 Two-Dimensional Electrophoresis

1. Focusing solution: 7 M urea, 2 M thiourea, 4 % CHAPS, 20 mM dithiothreitol (DTT) (BioRad) and 0.5 % v/v carrier Ampholytes cocktail (Bio-Lytes 3/10 Ampholytes, BioRad).
2. IPG strips, 18 cm, pH 4–7 (GE Healthcare).

3. Mineral oil (BioRad).
4. Equilibration buffer: 1.5 M Tris–HCl pH 8.8, 6 M urea, 30 % glycerol containing 10 mg/ml DTT (BioRad) or 25 mg/ml iodoacetamide (IAA) (BioRad).
5. 2-DE second dimension and SDS-PAGE: Acrylamide/Bisacrylamide (Proteogel), 1 M Tris–HCl pH 8.8, 1 M Tris–HCl pH 6.8, SDS (10 % solution), TEMED (BioRad), Ammonium Persulphate, APS (BioRad) (10 % solution, freshly prepared). Polyacrylamide gels of desired size and composition prepared according to standard protocols. Running buffer (1×): 0.03 M Tris, 0.2 M glycine, 0.1 % SDS.
6. Protean IEF Cell (BioRad) or IPGphor (GE Healthcare) for IEF step and Ettan Dalt Six (GE Healthcare) for the second dimension.
7. Silver staining kit, protein Plus One (GE Healthcare). Fixative: 40 % ethanol, 10 % acetic acid. Wash solution: Milli-Q water. Sensitizer: 30 % ethanol, 4 % sodium thiosulphate (5 % w/v solution), 68 mg/ml sodium acetate. Silver solution: 0.25 % silver nitrate. Developer: 25 mg/ml sodium carbonate with 20–40 μl formaldehyde (37 % w/v solution) per 100 ml total volume. Stopper: 14.6 mg/ml EDTA (*see* **Note 1**).

2.4 Proteins Digestion and Peptide Analysis by LC-MS/MS Analysis

1. DTT (BioRad), IAA (BioRad), modified porcine trypsin (Promega), trifluoroacetic acid, acetic acid, formic acid.
2. Mobile phases: (A) 0.5 % acetic acid and (B) 0.5 % acetic acid in 95 % acetonitrile or (A) 0.1 % formic acid and (B) 0.1 % formic acid in acetonitrile.
3. Columns: C-18 reversed phase nano-column (100 mm i.d. and 12 cm, Mediterranea Sea, Teknokroma), BioBasic C-18 PicoFrit column (75 mm, 10 cm; New Objective).
4. LTQ-Orbitrap XL mass spectrometer (Thermo Fisher, San José, CA, USA) and LTQ linear ion trap mass spectrometer (Thermo Electron, San Jose, CA).

3 Methods

3.1 Sample Collection and Red Blood Cells Isolation

1. Collect blood samples in EDTA tubes to prevent coagulation.
2. Place the tubes in N_2 inert atmosphere to prevent oxidation (*see* **Note 2**) at 4 °C for 3 days without shaking to allow maturation of reticulocytes to erythrocytes.
3. Shortly centrifuge (524 ×*g*, 10 min at 4 °C) to remove plasma (upper phase). A buffy layer containing white blood cells is observed in the interface between erythrocytes and plasma, which has to be removed (*see* **Note 3**).

4. Wash erythrocytes fraction by adding cold wash buffer up to 10–15 ml total volume and pipetting vigorously up and down to mix buffer with red cells extensively. Centrifuge again (524 × *g*, 10 min, 4 °C) and discard upper liquid phase. Repeat this wash step for a total of three times. For the last wash, transfer the red cells to ultracentrifugation tubes (*see* **Note 4**).

3.2 Fractionation of Erythrocyte Membrane and Cytosol

1. Perform hypotonic lysis by adding 13 ml of cold lysis buffer in continuous agitation. Centrifuge at 13,300 × *g* for 20 min at 4 °C. In this first lysis step, a whitish fluid pellet is formed and the upper liquid layer becomes red due to hemoglobin release from inner of the cells, which difficults pellet visibility (*see* **Note 5**). Keep this liquid upper layer as cytosolic fraction at −80 °C until analysis or proceed with hemoglobin depletion (*see* **Note 6**).
2. Repeat lysis step a total of five times to maximize removal of hemoglobin bound to membrane fraction. In the last cycle, centrifuge at 23,700 × *g* to collect membrane pellets (*see* **Notes 7** and **8**).
3. To further remove liquid layer, transfer membrane pellets to Eppendorf tubes and repeat centrifugation (20,000 × *g*, 15 min, 4 °C). Aliquot and store at −80 °C until analysis.

3.3 Cytosolic Hemoglobin Depletion and Sample Pretreatment Prior to 2-DE

Cytosolic fraction was depleted from hemoglobin by a capture reagent named HemogloBind™.

1. Shake the HemogloBind™ suspension and transfer 100 μl to an Eppendorf tube.
2. Add 100 μl of the isolated cytosolic fraction (*see* **Note 9**).
3. Vortex for 20 s and mix by inversion for 15 min. Centrifuge at 3,000 × *g* for 3 min and collect supernatant (*see* **Notes 10** and **11**).
4. Desalt in Protein Desalting Spin Column as follows (*see* **Note 12**):
   (a) Invert several times to resuspend slurry, twist off bottom closure, and loosen cap. Remove excess liquid by centrifugation at 1,500 × *g* for 1 min (a mark will help orientating the column in the same way for all successive centrifugation steps).
   (b) Empty the collection tube and replace column into it.
   (c) Add 300 μl of 2-DE sample buffer to the top of the column and do not mix. Centrifuge (1,500 × *g*, 1 min) to remove excess liquid. Repeat this wash steps twice, centrifuging for 2 min in the last one. Transfer column to a new tube.
   (d) Load up to 120 μl total volume per column of depleted cytosolic fraction onto the center of the compacted resin of

the desalting column avoiding disturbance of the resin or sample flow around.

(e) Centrifuge at 1,500 × *g* for 2 min and collect the desalted cytosolic proteins fraction, ready to analyze by 2-DE.

3.4 Membrane Fraction Pretreatment Prior to Analysis by 2-DE

In the present chapter we describe a protocol setup optimized in our laboratory that combines hypotonic lysis and purification steps of the "2-D Sample Prep for Membrane Proteins" commercial kit with modifications (11).

1. Preparation of the membrane extract.
   (a) Solubilize 200 μl erythrocytes membrane fraction (1:1) in lysis buffer for 30 min with occasional vortex and sonication.
   (b) Add 450 μl Mem-PER® solution and vortex.
   (c) Incubate tubes for 30 min on ice, vortexing every 5 min.
   (d) Centrifuge tubes at 10,000 × *g* for 3 min at 4 °C to remove cell debris. Transfer supernatant to new tubes and incubate for 10 min at 37 °C. Vortex and aliquot in 250 μl fractions (*see* **Note 13**).
   (e) Dilute 1 part Mem-PER® buffer with three parts ultrapure water and add 50 μl of this solution to each of the 250 μl aliquots. At this point, samples can be stored at −80 °C or proceed to the membrane extract purification step as detailed below.
2. Purification of membrane proteins.
   (a) Shake and vortex the 2-D PAGEprep™ resin to evenly disperse the resin and pipette 20 μl into a spin cup inserted in a collection tube (*see* **Note 14**).
   (b) Transfer all the lysed membrane fraction (300 μl) to the spin cup and vortex spin cup for 5 min to mix.
   (c) Transfer the same volume of 100 % DMSO to the spin cup to make a 1:1 ratio with the membrane extract. Vortex briefly to mix and incubate for 5 min at room temperature with occasional vortexing.
   (d) Centrifuge at 3,000 × *g* for 4 min (*see* **Note 15**).
   (e) Add 500 μl of 50 % DMSO solution to the resin. Vortex for 1 min.
   (f) Centrifuge (3,000 × *g*, 4 min), discard wash, and blot collection tube on a paper towel before reinserting spin cup into the same collection tube. Repeat for a total of three washes.
   (g) With a new collection tube, add 60 μl of elution buffer to the resin, vortex briefly, and incubate spin cup at 60 °C for 5 min. Vortex again and centrifuge at 3,000 × *g* for 3 min. Collect eluate on ice.

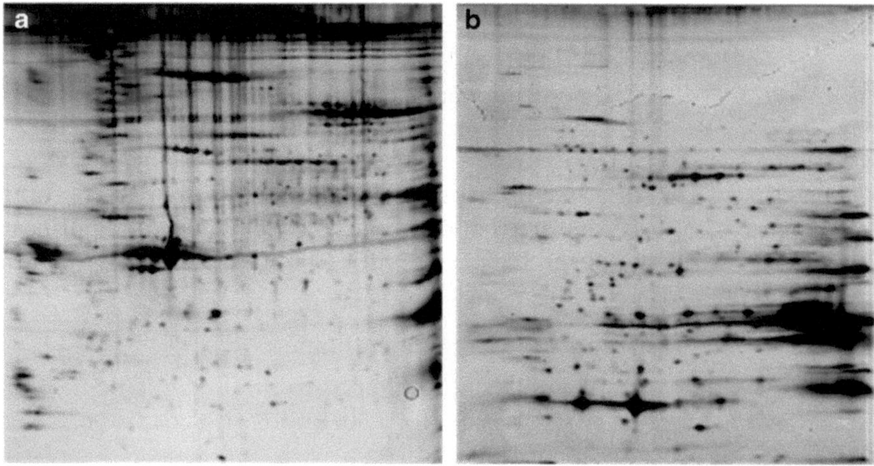

Fig. 1 Typical 2-DE patterns observed for erythrocytes membrane (**a**) and cytosolic (**b**) sub-proteomes

3. Desalting of membrane proteins prior to 2-DE

   Follow protocol as described for the cytosolic fraction. Prior to sample loading, centrifuge heavily (i.e., 8,000 × *g*, 2 min) the membrane extract that was eluted from the purification resin to remove any particulate material. Load all purified membrane fraction (about 60 μl) onto the center of the compacted resin.

3.5 Two-Dimensional Electrophoresis of Erythrocyte Membrane and Cytosolic Proteins

1. Solubilize purified membrane or cytosolic proteins fraction in the focusing solution to a final volume of 350 μl.
2. Spin at high speed for a few minutes to remove any undissolved material and load onto 18 cm length IPG pH 4–7 strips. Cover IPG strips with mineral oil to prevent evaporation.
3. Start IEF with active rehydration at 50 V for 12 h, from 50 to 300 V in 1 min; maintain at 300 V for 3 h, from 300 to 3,500 V in 1 h; and maintain at 3,500 V for 20 h.
4. After IEF, load strips onto an equilibration tray with the acrylamide surface upwards and equilibrate in two steps of 20 min, with gentle agitation in equilibration buffer containing DTT and IAA, respectively. Slightly wash IPG strips with Milli-Q water and load them on top of 10 % polymerized acrylamide ensuring no air bubbles remained trapped between the strip and the gel and with extreme caution to prevent strip acrylamide damage (*see* **Note 16**).
5. Run second dimension on Ettan Dalt Six chamber at maximum 13 W per gel with running buffer 1× overday (*see* **Note 17**) or at 1 W per gel overnight. Stop electrophoresis when the blue front reaches the bottom of the gel.
6. Silver staining of 2-DE gels is performed following manufacturer's protocols (i.e., GE Healthcare) (*see* **Note 1**) (Fig. 1).

3.6 nLC-MS/MS Membrane Proteins Identification

1. Once isolated by hypotonic lysis, mix the membrane fraction with buffer (1:1) (7 M urea, 2 M thiourea or 7 M urea, 2 M thiourea, 2 % ASB-14) for 30 min. Reduce with DTT (10 mM) and alkylate Cys groups with IAA (50 mM).
2. Perform digestion with modified porcine trypsin at a final ratio of 1:50 (trypsin: protein), overnight at 37 °C. Vacuum-dry the samples and dissolve in 1 % acetic acid for LC-MS analysis.
3. Inject the resulting tryptic peptides onto a C-18 reversed phase nano-column and analyze in a continuous acetonitrile gradient consisting of 0–50 % B in 45 min and 50–90 % B in 1 min (B: 0.5 % acetic acid in 95 % acetonitrile). Elute peptides at a flow rate of 300 nl/min from the RP nano-column to an emitter nanospray needle for real-time ionization and peptide fragmentation on an LTQ-Orbitrap mass spectrometer. Along the chromatographic run (70 min), analyze an enhanced FT-resolution spectrum (resolution 30,000) followed by the MS/MS spectra from the most intense five parent ions. Set dynamic exclusion at 1 min.
4. Search fragmentation spectra against MSDB database using the MASCOT (Swiss-Prot database) and SEQUEST (NCBI database) programs for protein identification and posttranslational modification characterization. Allow two mixed cleavages and set an error of 10 ppm or 0.8 Da for full MS or MS/MS spectra searches, respectively. Perform identifications by Proteome Discoverer 1.0 software (Thermo Fisher, San Jose, CA). Set decoy database search for false discovery rate (FDR) analysis at 0.05 by applying corresponding filters. A multiconsensus list can be created with results from two lysed aliquots.

3.7 nLC-MS/MS Cytosolic Proteins Identification

1. Deplete hemoglobin from a cytosolic fraction isolated by hypotonic lysis and precipitate proteins by TCA/acetone method. Solubilize the pellet in 0.5 % TFA, reduce, alkylate, and digest with trypsin.
2. Trap and desalt peptides in the trap column for 5 min and then separate on a BioBasic C-18 PicoFrit column. Use as LC solvents (A) 0.1 % formic acid in water and (B) 0.1 % formic acid in acetonitrile. Use a gradient separation of 5 % B for 5 min, 5–70 % B over 120 min, and 98 % B for 15 min at a flow rate of 200 nl/min. Analyze the eluted peptides in an LTQ linear ion trap mass spectrometer in data-dependent MS experiments using as parameters for ion scanning: full-scan MS (400–16 500 m/z) plus top five peaks Zoom/MS/MS (isolation width 2 m/z), normalized collision energy 35 %.
3. Use a dynamic exclusion list (30 s exclusion list size of 500 and exclusion width of 3 m/z) and Bioworks 3.2 software (Thermo Electron) to search the IPI human protein database (ipi. Human.v3.20.fasta, created on 03/01/2007, revision 12.2,

61 225 proteins) with peptide tolerance of 200 ppm, fragment ion tolerance of 0.5 Da, and 1 missed cleavage sites. Use a peptide probability of 0.001 and a crossing correlation score (Xcorr) of 2.0 (+2) and 2.5 (+3) as cutoff filters (due to this restrictive criteria, protein identifications based on a unique peptide can be also considered) (*see* **Note 18**).

4 Notes

1. A conventional protocol that works fine is fixing for a minimum of 30 min (12 % methanol and 7 % acetic acid is another option), sensitizing for 15–20 min (cold solution), three water washes for 5 min each, silver staining for 15 min, two water washes for 1 min (time should be carefully controlled), developing (check carefully to avoid over-staining) and stopping for 10 min. Gels can be stored in water once 2–3 washes have been performed to remove EDTA. Avoid glutaraldehyde in the sensitizing solution since it induces cross-linking of proteins, interfering with mass spectrometry.
2. This can be done by smooth bubbling of N_2 gas over the liquid surface, very carefully and without direct contact; tubes can be then capped and sealed with parafilm.
3. It is preferably to lose part of erythrocytes in favor of purity, so part of them is removed together with the buffy layer.
4. The supernatant should be clear; otherwise, hemolysis has occurred and samples have to be discarded. Residual buffy layer may also be observed in the first wash and has to be discarded.
5. Whole red cells produce a cloudy, opaque suspension but once the cells are lysed the suspension turns clear, although dark red. Lysis releases the red cell contents (almost entirely hemoglobin) into the buffer, leaving empty membrane sacks or red cell "ghosts" in suspension.
6. At this stage, it is preferably to leave about half milliliter of cytosol together with the pellet than trying to keep all cytosolic volume at risk of simultaneously removing membrane pellet; due to its low viscosity and the red color of the liquid phase, there is a risk of sucking the membranes up as the supernatant is collected.
7. In the successive lysis steps, the membrane pellet is being washed to remove most of the remaining hemoglobin, and so the supernatant becomes lighter in color and the membrane pellet more visible.
8. Pay particular attention to the small "button" of material that may appear near the bottom of the tube, as it is a tangle of fibrin, white cells, platelets, and unlysed red cells which should

not be confused with the membrane pellet and therefore should be discarded.

9. Different ratios of sample:depletion reagent were tested (i.e., 100, 80, and 40 μl (two times) HemogloBind™ reagent for 100 μl cytosolic fraction) and no significant differences were observed (11).
10. Even if the solution is not totally colorless, hemoglobin depletion guarantees 2-DE interference-free profiles.
11. Analysis of proteins co-retained with hemoglobin by SDS-PAGE and mass spectrometry revealed hemoglobin isoforms and other minority proteins as catalase, G3PDH, carbonic anhydrase 1 and 2, flavin reductase, and peroxiredoxin 2 (11).
12. The HemogloBind™ reagent is totally incompatible with direct load of depleted sample in 2-DE gels, causing severe streaking.
13. The hydrophobic (membrane) and hydrophilic (cytosolic) fractions have been previously separated by hypotonic lysis, so the phase's partition referred in the kit protocol is not visible; in any case, this additional lysis is applied to have membrane proteins in adequate buffer for further steps.
14. The slurry is viscous and will require a cut pipette tip.
15. Sometimes the solid gets so tight that it may impede proper flow through; in that case, repeat the centrifugation step and if it does not work, recover the liquid by pipetting out.
16. Once loaded onto the gel, IPG strips can be sealed by agarose solution (0.5 %) preheated at 60–90 °C and loaded all over the strip once on top of the gel surface; once cold agarose solution becomes solid, IPG strip is immobilized while empty spaces are filled in, ensuring optimum transfer of proteins into the gel.
17. Upper chamber should be better filled with running buffer 2× to facilitate migration of high-molecular-weight proteins into the gel.
18. These nLC-MS/MS sections show one of the multiple possibilities for proteins identification by mass spectrometry of the erythrocyte sub-proteomes.

Acknowledgments

This work was supported by Instituto de Salud Carlos III (grant numbers FIS PI070537, FIS PI080970, FISPI080920, RD07/0064/0023, FI06/00583, CP09/00229, PI11/02239, PI11/001401), Mutua Madrileña Automovilista (grant number 20174/004), Fundacion Conchita Rabago, Redes temáticas de Investigación Cooperativa en Salud (RD06/0014/1015), and grants from Fundación para la Investigación Sanitaria de Castilla-La Mancha (FISCAM PI2008-08, PI2008-28, PI2008-52).

References

1. Goodman SR, Kurdia A, Ammann L, Kakhniashvili D, Daescu O (2007) The human red blood cell proteome and interactome. Exp Biol Med 232:1391–1408
2. Kakhniashvili DG, Bulla LA Jr, Goodman SR (2004) The human erythrocyte proteome: analysis by ion trap mass spectrometry. Mol Cell Proteomics 3:501–509
3. Pasini EM, Kirkegaard M, Mortensen P, Lutz HU et al (2006) In-depth analysis of the membrane and cytosolic proteome of red blood cells. Blood 108:791–801
4. Zhang Q, Tang N, Schepmoes AA, Phillips LS et al (2008) Proteomic profiling of no enzymatically glycated proteins in human plasma erythrocyte membranes. J Proteome Res 7:2025–2032
5. Santoni V, Mohillo M, Rabilloud T (2000) Membrane proteins and proteomics: un amour impossible? Electrophoresis 21:1054–1070
6. Prabakaran S, Wengenroth M, Lockstone HE, Lilley K et al (2007) 2-D DIGE analysis of liver and red blood cells provides further evidence for oxidative stress in schizophrenia. J Proteome Res 6:141–149
7. Jiang M, Jia L, Jiang W, Hu X et al (2003) Protein dysregulation in red blood cell membranes of type 2 diabetic patients. Biochem Biophys Res Commun 309:196–200
8. Kakhniashvili DG, Griko NB, Bulla LA Jr, Goodman SR (2005) The proteomics of sickle cell disease: profiling of erythrocyte membrane proteins by 2D-DIGE and tandem mass spectrometry. Exp Biol Med 230:787–792
9. Luche S, Santoni V, Rabilloud T (2003) Evaluation of nonionic zwitterionic detergents as membrane protein solubilizers in two-dimensional electrophoresis. Proteomics 3: 249–253
10. Bruschi M, Seppi C, Arena S, Musante L et al (2005) Proteomic analysis of erythrocyte membranes by soft Immobiline gels combined with differential protein extraction. J Proteome Res 4:1304–1309
11. Alvarez-Llamas G, de la Cuesta F, Barderas MG, Darde VM, Zubiri I, Caramelo C, Vivanco F (2009) A novel methodology for the analysis of membrane and cytosolic sub-proteomes of erythrocytes by 2-DE. Electrophoresis 23: 4095–4108

Chapter 7

Characterization and Analysis of Human Arterial Tissue Secretome by 2-DE and nLC-MS/MS

Fernando de la Cuesta, Maria G. Barderas, Enrique Calvo, Irene Zubiri, Aroa S. Maroto, Juan Antonio Lopez, Fernando Vivanco, and Gloria Alvarez-Llamas

Abstract

Early detection of cardiovascular diseases and knowledge of underlying mechanisms is essential. Tissue secretome studies resemble more closely to the in vivo situation, showing a much narrower protein concentrations dynamic range than plasma. In the present chapter, we detail the characterization and analysis of human arterial tissue secretome by two-dimensional electrophoresis (2-DE) and nano-liquid chromatography on-line coupled to mass spectrometry (nLC-MS/MS). General strategies shown here can be extended to other tissue secretome studies.

Key words Secretome, Arterial tissue, Coronary artery, Mammary artery, Atherosclerosis, 2-DE, nLC-MS/MS, Label-free quantitation

1 Introduction

The secretome is the subset of proteins released by a cell/tissue under certain conditions. One of its main advantages lies on the fact that, if properly obtained, it shows a much narrower protein concentrations dynamic range (reduced complexity) than serum or plasma. Particularly, studies focused on tissue secretome resemble more closely to the in vivo situation compared to cell culture workflows. Working directly with tissue in culture instead of cells better simulates the in vivo condition, and secretory molecules coming from all tissue components can be thus detected as a result of cross-talk between them. Optimum culture conditions for tissue, in general, and arterial tissue, in particular, have to be carefully investigated in the aim to reduce plasma contamination and detect low-abundance secreted proteins (1). In this sense, a culture protocol based on several washing steps (culture media exchange) adequately distributed in time minimizes plasma contamination, resulting in a reduced

Fernando Vivanco (ed.), *Vascular Proteomics: Methods and Protocols*, Methods in Molecular Biology, vol. 1000, DOI 10.1007/978-1-62703-405-0_7,

protein concentration dynamic range and increasing the possibilities to detect low-abundance secreted proteins. Cardiovascular diseases are the number one cause of death globally. However, operating mechanisms are not fully understood. By studying the secretome from preatherosclerotic and atherosclerotic arterial tissue, novel acting proteins will be discovered and mechanisms of action will be better understood (2–4). Complications derived from rupture or damage in coronary plaque are one of the main causes of death, but the reduced diameter of coronary artery and the obtaining difficulties have limited the amount of available material. Despite of being a main target with direct implication in atherosclerosis, a scarce number of studies have been focused on human coronary arterial tissue (5–8). We detail here a protocol for extensive analysis of human coronary and mammary arterial tissue which provides with a wide picture of released proteins. Label-free quantitative analysis can additionally be performed to find out differentially expressed proteins between healthy and pathological conditions (9).

2 Materials

2.1 Tissue Culture

1. PBS: phosphate saline buffer, pH 7.2.
2. Culture medium: 1640 RPMI medium without l-Glutamine (GIBCO, Invitrogen) supplemented with penicillin, streptomycin (Pen/Strep, Lonza) (100 units/ml), and 0.05 % amphotericin B (Fungizone 50 mg, Bristol-Myers Squibb) (2.5 μg/ml).
3. Twenty-four-well tissue culture plate (Iwaki).
4. Sterile tweezers and surgical blades (Braun).

2.2 Histology

1. OCT (optimal cutting temperature) compound (Tissue-Tek, Sakura).
2. Antibodies anti-smooth muscle actin (Clone 1A4, 1:500) and anti-CD68 (Clone PG-M1, dilution1:500) from Dako.
3. Secondary antibody: peroxidase-conjugated EnVision + Dual Link (Ready-to-use solution, Dako).
4. Mayer's hematoxylin solution and eosin Y alcoholic solution (Sigma-Aldrich), xylol, ethanol, DPX nonaqueous mounting medium for microscopy (Merck).
5. Hydrogen peroxide (3 %), BSA (10 % in EnVision wash buffer, Dako), liquid DAB+ substrate chromogen system (Dako), ethanol (70 %, 90 %, 100 % solutions).
6. Wash buffer: EnVision wash buffer, Dako.

2.3 Gel Electrophoresis (2-DE and SDS-PAGE)

1. Amicon Ultra 5 kDa cutoff (Millipore).
2. Rehydration buffer: 7 M urea (BioRad), 2 M thiourea (BioRad), 4 % CHAPS (BioRad), 0.6 % Ampholites (BioRad), 0.6 % TBP

(BioRad), 1.2 % DeStreak reagent (GE Healthcare) and blue bromophenol (BioRad).

3. Mineral oil (BioRad).
4. IPG strips 4–7, 18–24 cm (GE Healthcare).
5. Equilibration buffer: 1.5 M Tris–HCl pH 8.8, 6 M urea, 30 % glycerol containing 10 mg/ml DTT (BioRad) or 25 mg/ml IAA (BioRad).
6. 2-DE second dimension and SDS-PAGE: acrylamide/Bisacrylamide (Proteogel), 1 M Tris–HCl pH 8.8, 1 M Tris–HCl pH 6.8, SDS (10 % solution), TEMED (BioRad), APS (BioRad) (10 % solution, freshly prepared). Polyacrylamide gels of desired size and composition prepared according to standard protocols. Running buffer (1×): 0.03 M Tris, 0.2 M glycine, 0.1 % SDS.
7. Coomassie Brilliant Blue G-250 for gel staining (prior to nLC-MS/MS).
8. Protean IEF Cell (BioRad) or IPGphor (GE Healthcare) for IEF step. Ettan Dalt Six (GE Healthcare) and MiniProtean Tetra Cell (BioRad) for SDS-PAGE.

2.4 Silver Staining

Most components are part of the kit "Silver staining kit, protein Plus One" (GE Healthcare) (*see* **Note 1**).

1. Fixative: 40 % ethanol, 10 % acetic acid.
2. Wash solution: Milli-Q water.
3. Sensitizer: 30 % ethanol, 4 % sodium thiosulphate (5 % w/v solution), 68 mg/ml sodium acetate.
4. Silver solution: 0.25 % silver nitrate.
5. Developer: 25 mg/ml sodium carbonate, add 20–40 µl formaldehyde (37 % w/v solution) per 100 ml total volume.
6. Stopper: 14.6 mg/ml EDTA.

2.5 Proteins Digestion and Peptide Analysis by LC-MS/MS Analysis

1. Digestion in gel: DTT (10 mM, BioRad), iodoacetamide (50 mM, BioRad), modified porcine trypsin (Promega), trifluoroacetic acid (0.5 %, Merck), acetic acid (1 %, Merck).
2. Mobile phases: (a) 0.5 % acetic acid and (b) 95 % acetonitrile, 0.5 % acetic acid.
3. Columns: trapping cartridge (RP C-18), C-18 reversed-phase (RP) nano-column (100 µm I.D. and 12 cm, Mediterranea Sea, Teknokroma).
4. PicoTip™ emitter nano-spray needle (New Objective, Woburn, MA).
5. LTQ-Orbitrap XL mass spectrometer (Thermo Fisher, San José, CA, USA).

6. MS/MS analysis: Sequest (Thermo Fisher Scientific; version 1.0.43.2) and X! Tandem (The GPM, thegpm.org; version 2007.01.01.1).
7. Label-free semiquantitation: Scaffold software (Proteome Software Inc., Portland, OR).

3 Methods

3.1 Arterial Tissue Collection and Culture

1. Once obtained from surgery/autopsy, transport arterial tissue immediately to the laboratory on ice and immerse in PBS (*see* **Note 2**).
2. Wash tissue samples extensively and several times in PBS and culture medium to eliminate major serum contaminants.
3. Over a clean surface area (i.e., Petri dish), remove blood and fat deposits from the arterial tissue surface with tweezers and clean it with saline solution as much as possible.
4. Section tissue into pieces of about 2–3 mm length and transfer them to a 24-well tissue culture plate for culture. Place 70–100 mg arterial tissue per dish in 1 ml culture medium and incubate at 37 °C in a humidified atmosphere of 5 % CO_2.
5. Replace culture medium after 1 h, overnight, and every 4 h in the following 8 h period (a total of four washes) (*see* **Note 3**).
6. Maintain tissue in culture for additional 48 h prior to collection of final secretome samples which can be stored at −80 °C until analysis.

3.2 Arterial Tissue Histology (Fig. 1)

1. Hematoxylin and eosin staining: fix in 70 % ethanol; wash with water to remove OCT; stain in hematoxylin solution for 5–10 min; wash in running tap water; dip two to three times in eosin solution; dehydrate for successive immersions in ethanol 70 %, 95 %, and 100 % solutions; clear in xylol; and mount in DPX.
2. Actin staining: fix in acetone (−20 °C, 5 min); wash with water to remove OCT; mark selected area with immunopen; block with BSA (10 %, 1 h); incubate with smooth muscle actin primary antibody (30 min); wash with wash buffer; remove peroxidases with hydrogen peroxide (3 %, 5 min); wash with wash buffer; incubate with secondary antibody (30 min); wash with wash buffer; incubate with DAB (5–10 min); wash with water; stain in hematoxylin solution (30 s–1 min; *see* **Note 4**): wash in running tap water; dehydrate by successive immersions in ethanol 70 %, 90 %, and 100 % solutions; clear in xylol; and mount in DPX.
3. CD68 staining: follow actin staining protocol with anti-CD68 as primary antibody (*see* **Note 5**).

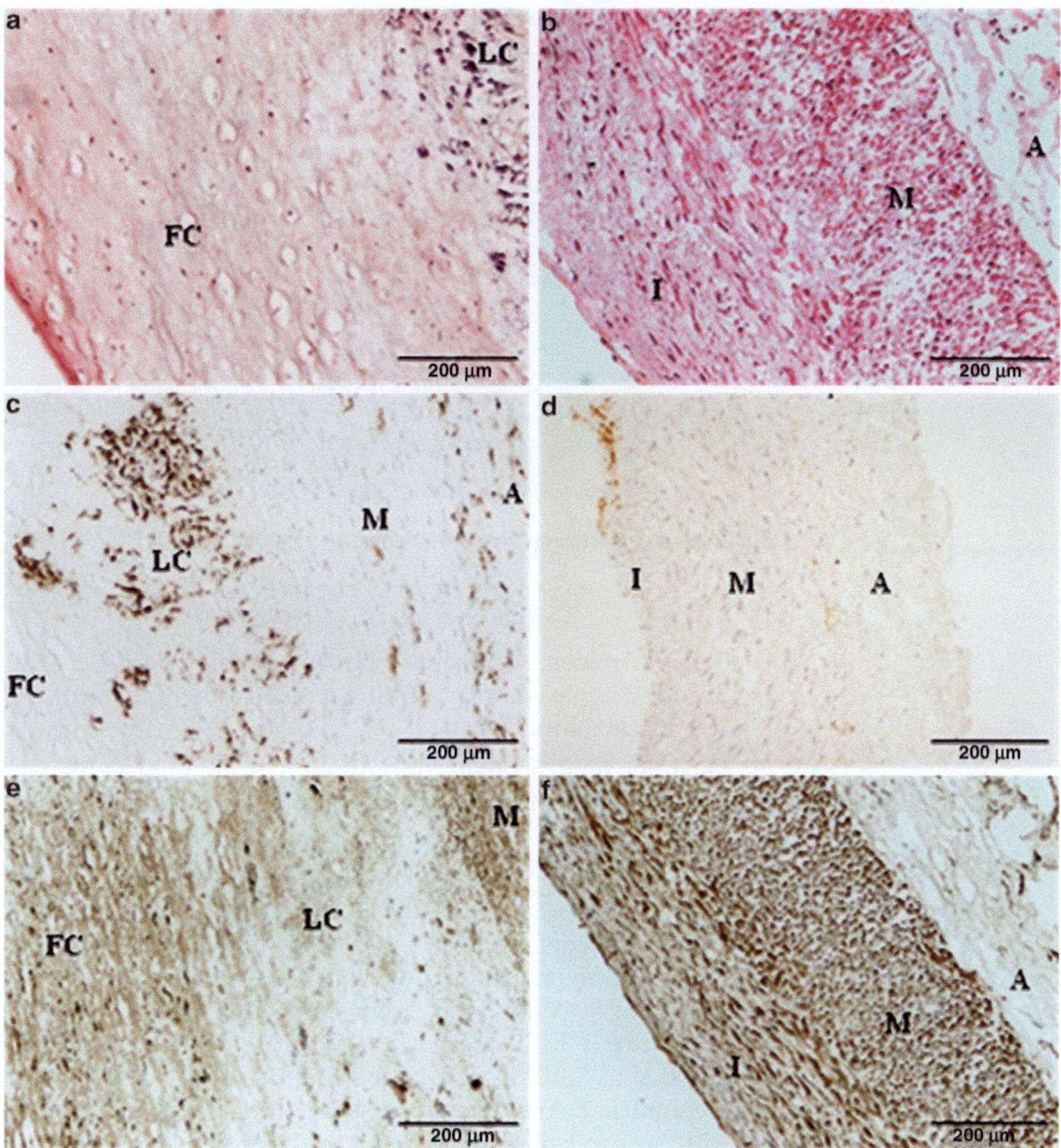

Fig. 1 Histological characterization of human artery material. Arterial tissue was stained with H&E. Atherosclerotic coronary (AC) showing a lipid core surrounded by a fibrous cap (**a**) and preatherosclerotic (PC) showing incipient intimal thickening (**b**) are shown. Immunohistochemistry analysis of CD68 antigen allowed colocalizing this lipid deposition with macrophage/foam cells staining both in AC (**c**) and mammary artery (**d**). After actin immunostaining, intimal smooth muscle cells (SMCs) were found to be mainly located in the fibrous cap of the AC (**e**). Abundant subendothelial SMCs were observed within the PC intima, which are responsible for the intimal thickening (**f**). *FC* fibrous cap, *LC* lipid core, *I* intima, *M* media, and *A* adventitia (From (9) with permissions)

3.3 Secretome Analysis by 2-DE

1. Concentrate proteins by ultrafiltration (*see* **Note 6**). Collect the 0.7–1 ml media remaining in the dish after culture period, centrifuge at 20,000 × *g* for 15 min (to remove traces of tissue pellet), and filter supernatant through 5 kDa cutoff ultrafiltration

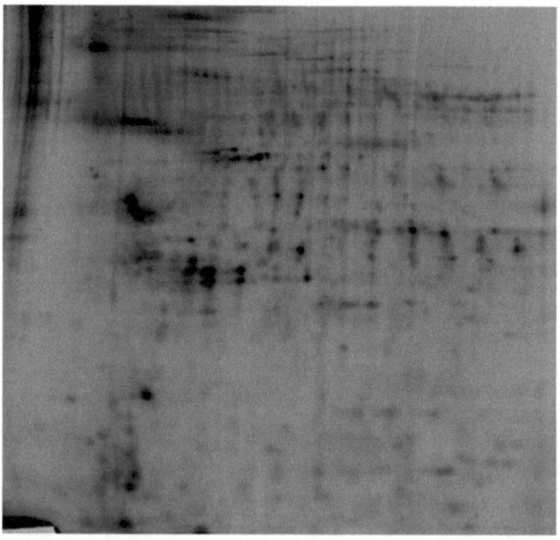

Fig. 2 Secretome from coronary arterial tissue analyzed by 2-DE

device (Amicon, Millipore). Spin at maximum 7,500 g with a 35° fixed angle rotor or 4,000 g with a swinging bucket rotor at 4 °C (*see* **Note** 7).

2. Quantify total protein concentration by Bradford, BCA, or similar methodologies for that purpose.
3. Mix appropriate sample volume (i.e., 100–200 μg total protein) with rehydration buffer, spin at high speed for a few minutes to remove any undissolved material, and load onto, i.e., 18–24 cm length IPG pH 4–7 strips. Cover IPG strips with mineral oil to prevent evaporation. IEF starts with active rehydration at 50 V for 12 h, followed by 200 V for 1 h 30 min, 500 V for 1 h 30 min, 1,000 V for 1 h 30 min, and from 1,000 V to 8,000 V in 1 h, and 8,000 V until 50,000 V were accumulated. After IEF, load strips onto an equilibration tray with the acrylamide surface upwards.
4. Equilibrate in two steps of 20 min with gentle agitation in equilibration buffer containing DTT and IAA, respectively. Slightly wash IPG strips with Milli-Q water and load them on top of 12 % polymerized acrylamide ensuring no air bubbles remained trapped between the strip and the gel and with extreme caution to prevent strip acrylamide damage (*see* **Note 8**).
5. Run second dimension on Ettan Dalt Six chamber at maximum 13 W per gel (18–24 cm IPG strips) with running buffer 1× overday (*see* **Note 9**) or at 1 W per gel overnight. Stop electrophoresis when the blue front reaches the bottom of the gel.
6. Silver staining of 2-DE gels is performed following manufacturer's protocols (i.e., GE Healthcare) (*see* **Note 1**) (Fig. 2).

3.4 Secretome Analysis by nLC-MS/MS

3.4.1 Proteins Digestion and Peptides Extraction

1. Load 30 μg total protein secretome sample (previously concentrated by ultrafiltration) on a self-poured stacking SDS-PAGE gel (10 % resolving gel and 4 % stacking gel) at 20 mA/gel.
2. Stop the electrophoresis when the front dye barely passes from the stacking gel into the resolving gel, before separation of protein mixture into discrete bands along the gel takes place. In this way, all proteins concentrate in a unique band, eliminating sample contaminants and facilitating reproducibility for comparisons.
3. Stain the unique concentrated band obtained for every sample by Coomassie Brilliant Blue G-250.
4. Excise the band and cut it into small pieces. Store at 4 °C in ultrapure water until analysis.
5. Proteins digestion and peptides extraction. Follow Shevchenko et al.'s protocol with minor variations (10): reduce disulphide bonds with DTT (10 mM), alkylate cysteine groups with iodoacetamide (50 mM), add modified porcine trypsin (Promega) at a final ratio of 1:50 (trypsin-protein), and perform tryptic digestion overnight at 37 °C. Extract tryptic peptides with 0.5 % trifluoroacetic acid, followed by vacuum drying. Dissolve in 20 μL 1 % acetic acid for identification by nLC-MS/MS analysis.

3.4.2 Analysis by nLC-MS/MS

1. Use a trapping cartridge to on-line desalt peptides and separate them onto a C-18 reversed-phase nano-column in a continuous acetonitrile gradient consisting of 0–43 % B in 140 min and 50–90 % B in 1 min with a flow rate of 300 nL/min.
2. Transfer eluted peptides from the RP nano-column to a PicoTip™ emitter nano-spray needle for real-time ionization and peptide fragmentation on an LTQ-Orbitrap XL mass spectrometer.
3. Typical instrument settings along the chromatographic run (180 min): enhanced FT-resolution spectrum (resolution = 60,000) followed by the MS/MS spectra from most intense five parent ions. Dynamic exclusion: 0.5 min. Extract tandem mass spectra and deconvolute charge state by Proteome Discoverer version 1.0 (Thermo Fisher Scientific). Analyze all MS/MS samples using Sequest (Thermo Fisher Scientific; version 1.0.43.2) and X! Tandem (The GPM, thegpm.org; version 2007.01.01.1). Set up X! Tandem to search a subset of the human database assuming the digestion enzyme trypsin. Set up Sequest to search human_ref.fasta. Fragment ion mass tolerance, 0.80 Da; parent ion tolerance, 10.0 ppm; and variable modifications: oxidation of methionine, acetylation of lysine, and phosphorylation of serine, threonine, and tyrosine.

4. Label-free semiquantitative comparison of different samples: Scaffold software (version Scaffold_3_00_03) (11). Establish protein and peptide probability greater than 99 % and 95 %, respectively. Compare proteins abundance between secretome samples based on unweighted spectral counts (method of counting peptides in all instances when they are shared between proteins) (*see* **Note 10**) (Table 1).

4 Notes

1. A conventional protocol that works fine is fixing for a minimum of 30 min (12 % methanol and 7 % acetic acid is another option), sensitizing for 15–20 min (cold solution), three water washes for 5 min each, silver staining for 15 min, two water washes for 1 min (time should be carefully controlled), developing (check carefully to avoid over-staining) and stopping for 10 min. Gels can be stored in water once 2–3 washes have been performed to remove EDTA. Avoid glutaraldehyde in the sensitizing solution since it induces cross-linking of proteins, interfering with mass spectrometry.
2. All material to be in contact with tissue has to be sterile. Maximize work in a laminar flow hood.
3. If the culture protocol is followed by SDS-PAGE or 2-DE analysis of successive removed wash media, progressive diminution of bands/spots corresponding to albumin and IgG can be seen, proving diminishing of contaminant plasma proteins; in parallel, enrichment of minor abundance proteins is achieved, resulting in a higher number of visible bands/spots that otherwise could have not been detected.
4. Hematoxylin solution can be reused and staining time is dependent of number of uses.
5. If not specified, water wash refers to MiliQ-water.
6. By ultrafiltration in Amicon devices, secretome sample volume can be reduced and proteins are thus concentrated.
7. Usually, 2,500 g for 2 h reduces volume from 0.7–1 ml to 0.1–0.2 ml.
8. Once loaded onto the gel, IPG strips can be sealed by agarose solution (0.5 %) preheated at 60–90 °C and loaded all over the strip once on top of the gel surface; once cold agarose solution becomes solid, IPG strip is immobilized while empty spaces are filled in, ensuring optimum transfer of proteins into the gel.
9. Upper chamber should be better filled with running buffer 2× to facilitate migration of high-molecular-weight proteins into the gel.
10. Do not use normalized data when the same amount of total protein was loaded in the mass spectrometer per sample.

Table 1
Proteins differentially released according to Scaffold software based on spectrum counts

Protein name	Accession number	Unweighted spectrum counts (av. per group)			Standard deviation			*t*-Test			Ratios		
		M	PC	AC	M	PC	AC	M-PC	M-AC	PC-AC	M-PC	M-AC	PC-AC
Vinculin	gi|4507877	37	19	15	4	14	7		0.006		2.0	2.5	1.3
Gelsolin	gi|4504165	33	16	23	7	9	13	0.012			2.0	1.4	0.7
Vimentin	gi|62414289	31	9	13	16	8	14	0.048	0.011		3.3	2.4	0.7
Lumican	gi|4505047	12	8	19	10	7	12		0.031		1.6	0.6	0.4
Phosphoglucomutase 5	gi|11415050	22	7	5	9	1	4		0.031		3.3	4.6	1.4
Lamin A/C isoform 1	gi|27436946	12	2	1	4	3	1		0.026		5.8	11.7	2.0
Albumin	gi|4502027	6	1	6	9	1	2			0.026	6.3	1.0	0.2
P3ECSL	gi|11545918	3	1	3	1	1	2	1.9×10^{-17}			4.0	0.8	0.2
Transgelin	gi|48255905	4	0	1	1	1	2	0.020			13.0	4.3	0.3
Similar to glutamate dehydrogenase 2, mitochondrial	gi|29745442	3	0	1	1	0	1	0.038			–	5.0	0.0
Peroxiredoxin 3 isoform b	gi|32483377	1	0	1	1	1	1	1.9×10^{-17}			4.0	2.0	0.5

M mammary, *PC* preatherosclerotic coronary, and *AC* coronary with atherome plaque

Acknowledgments

This work was supported by Instituto de Salud Carlos III (grant numbers FIS PI070537, FIS PI080970, FISPI080920, RD07/0064/0023, FI06/00583, CP09/00229, PI11/02239, PI11/001401), Mutua Madrileña Automovilista (grant number 20174/004), Fundacion Conchita Rabago, Redes Temáticas de Investigación Cooperativa en Salud (RD06/0014/1015), and grants from Fundación para la Investigación Sanitaria de Castilla-La Mancha (FISCAM PI2008-08, PI2008-28, PI2008-52).

References

1. Alvarez-Llamas G, Szalowska E, de Vries MP, Weening D, Landman K, Hoek A et al (2007) Characterization of the human visceral adipose tissue secretome. Mol Cell Proteomics 6:589–600
2. Duran MC, Mas S, Martin-Ventura JL, Meilhac O, Michel JB, Gallego-Delgado J et al (2003) Proteomic analysis of human vessels: application to atherosclerotic plaques. Proteomics 3:973–978
3. Martin-Ventura JL, Duran MC, Blanco-Colio LM, Meilhac O, Leclercq A, Michel JB et al (2004) Identification by a differential proteomic approach of heat shock protein 27 as a potential marker of atherosclerosis. Circulation 110:2216–2219
4. Martin-Ventura JL, Tunon J, Duran MC, Blanco-Colio LM, Vivanco F, Egido J (2006) Vascular protection of dual therapy (atorvastatin-amlodipine) in hypertensive patients. J Am Soc Nephrol 17:S189–S193
5. De la Cuesta F, Alvarez-Llamas G, Maroto AS, Zubiri I, Posada M, Padial LR, Barderas MG, Vivanco F (2011) A proteomic focus on the alterations occurring at the human atherosclerotic coronary intima. Mol Cell Proteomics 10:1–13
6. You SA, Archacki SR, Angheloiu G, Moravec CS, Rao S, Kinter M et al (2003) Proteomic approach to coronary atherosclerosis shows ferritin light chain as a significant marker: evidence consistent with iron hypothesis in atherosclerosis. Physiol Genom 13:25–30
7. Bagnato C, Thumar J, Mayya V, Hwang SI, Zebroski H, Claffey KP et al (2007) Proteomics analysis of human coronary atherosclerotic plaque: a feasibility study of direct tissue proteomics by liquid chromatography and tandem mass spectrometry. Mol Cell Proteomics 6:1088–1102
8. De la Cuesta F, Alvarez-Llamas G, Maroto AS, Donado A, Juarez-Tosina R, Rodriguez-Padial L et al (2009) An optimum method designed for 2-D DIGE analysis of human arterial intima and media layers isolated by laser microdissection. Proteomics Clin Appl 3:1174–1184
9. De la Cuesta F, Barderas MG, Calvo E, Zubiri I, Maroto AS, Darde VM et al (2012) Secretome analysis of atherosclerotic and non-atherosclerotic arteries reveals dynamic extracellular remodeling during pathogenesis. J Proteomics 75:2960–2971
10. Shevchenko A, Wilm M, Vorm O, Mann M (1996) Mass spectrometric sequencing of proteins silver-stained polyacrylamide gels. Anal Chem 68:850–858
11. Neilson KA, Ali NA, Muralidharan S, Mirzaei M, Mariani M, Assadourian G et al (2011) Less label, more free: approaches in label-free quantitative mass spectrometry. Proteomics 11: 535–553

Chapter 8

Identification of Novel Biomarkers of Abdominal Aortic Aneurysms by 2D-DIGE and MALDI-MS from AAA-Thrombus-Conditioned Media

Roxana Martinez-Pinna, Juan Antonio Lopez, Priscila Ramos-Mozo, Luis M. Blanco-Colio, Emilio Camafeita, Enrique Calvo, Olivier Meilhac, Jean Baptiste Michel, Jesús Egido, and José Luis Martin-Ventura

Abstract

In the search for novel biomarkers, noncandidate-based proteomic strategies open up new opportunities to gain a deeper insight into disease processes regarding their molecular mechanisms, the risk factors involved, and the monitoring of disease progression. To carry out these complex analyses, the combined use of gel electrophoresis with mass spectrometry (MS) represents a powerful choice. In addition, the introduction of protein dye labeling has notably improved the reliability of differential expression studies by increasing the statistical significance of the protein candidates. Here, we describe a strategy where different layers (luminal/abluminal) from the intraluminal thrombus (ILT) of human abdominal aortic aneurysm (AAA) patients were incubated in protein-free medium. Then, the levels of the proteins released were compared by two-dimensional differential in-gel electrophoresis (2D-DIGE) and the proteins of interest identified by MS. We consider that the use of tissue-conditioned media could offer a substantial advantage in the analytical study of biological fluids, as they provide a source of proteins to be released to the bloodstream, which could serve as potential circulating biomarkers.

Key words Differential proteomics, 2D-DIGE, Biomarkers, Conditioned media, Abdominal aortic aneurysms

1 Introduction

The scarce correlation between mRNA and protein levels, together with the complexity added by alternative splicing mechanisms and posttranslational modifications, points out that genome alteration alone fails to reflect the whole functional complexity of an organism. In this regard, proteomic approaches can provide essential information in different physiological scenarios, generally combining electrophoretic or chromatographic separations with mass spectrometry for protein identification (1–3). The past few years have witnessed tremendous advances in the development of

Fernando Vivanco (ed.), *Vascular Proteomics: Methods and Protocols*, Methods in Molecular Biology, vol. 1000, DOI 10.1007/978-1-62703-405-0_8, © Springer Science+Business Media New York 2013

proteomic tools, especially those related to MS-based approaches. Nevertheless, gel-based strategies continue to offer outstanding capabilities for quantitative proteomic studies. Thus, the 2D-DIGE system enables to separate differentially dye-labeled proteins in the same gel, therefore minimizing technical variability. The overlay of the corresponding images enables an accurate comparison of protein components among the samples, which provides a reliable quantitation of differential protein expression.

The complexity of vascular lesions, as well as that of biological fluids involved (e.g., plasma), hinders proteomic approaches. We hypothesized that the blood compartment could reflect what was observed in the arterial-conditioned medium and, thus, that proteins potentially released into circulation could serve as disease biomarkers. Our study, performed on the thrombus-conditioned medium, using 2D-DIGE labeling followed by MS, permitted the identification of several proteins associated to key mechanisms involved in AAA pathogenesis (4).

2 Materials

2.1 AAA-Tissue-Conditioned Media Preparation

1. Wash buffer: sterile saline buffer.
2. Culture medium: RPMI 1640 without fetal bovine serum, containing 1 % antibiotics and antimycotic (Gibco).

2.2 Sample Preparation for 2D-DIGE

1. 2D clean-up kit for protein precipitation (GE Healthcare).
2. Lysis buffer: 30 mM Tris–HCl pH 8.5, 7 M urea, 2 M thiourea, 4 % (w/v) CHAPS.
3. pH indicator paper strips pH 8.0–9.7 (Whatman International Ltd.).
4. RC-DC protein assay kit (Bio-Rad).

2.3 Protein Labeling for 2D-DIGE

1. Ten millimolar of L-Lysine monohydrochloride min 98 % (MW 182.6, Sigma).
2. Ultrapure anhydrous dimethylformamide (DMF) 99.8 % (DMF, Sigma). Stable for 3 months at room temperature.
3. Five nanomoles of CyDye DIGE Fluor Cy2 minimal dye, 5 nmol CyDye DIGE Fluor Cy3 minimal dyes, and 5 nmol CyDye DIGE Fluor Cy5 minimal dye (GE Healthcare). Store at −80 °C.

2.4 Sample Preparation and Isoelectric Focusing

1. Rehydratation buffer for sample loading: 7 M urea, 2 M thiourea, 4 % (w/v) CHAPS, 0.8 % (v/v) IPG Buffer 3-11NL, 50 mM DTT and bromophenol blue.

2. Rehydration buffer for Immobiline DryStrips: 7 M urea, 2 M thiourea, 4 % (w/v) CHAPS, 0.8 % (v/v) IPG Buffer 3-11NL, 97 mM DeStreak reagent (GE Healthcare) and bromophenol blue.
3. Immobiline DryStrip pH 3–11 NL, 24 cm and Immobiline DryStrip reswelling tray (GE Healthcare).
4. Paper wicks.
5. PlusOne™ DryStrip Cover Fluid (GE Healthcare).
6. IPGphor II IEF system (GE Healthcare).

2.5 Second Dimension (SDS-PAGE)

1. SDS equilibration buffer stock solution: 50 mM Tris–HCl, pH 8.0, 6 M urea, 40 % (w/v) glycerol and 1 % (w/v) SDS. This stock solution must be stored at −20 °C.
2. Reducing equilibration solution: 1 % (w/v) DTT in 10 mL SDS equilibration buffer stock solution. Use immediately after preparation.
3. Alkylating equilibration solution: 3.5 % (w/v) iodoacetamide in 10 mL SDS equilibration buffer stock solution. Use immediately after preparation.
4. Ettan DALTsix Large Vertical System for 24 cm strips, low-fluorescence plates with 1-mm integral spacers for Ettan DALT, Ettan DALT Cassette Rack, and Ettan DALT Gel Caster (all from GE Healthcare).
5. Water-saturated butanol. Use to overlay gels. Store at room temperature.
6. Two-dimensional DIGE SDS-PAGE (sodium dodecyl sulphate-polyacrylamide gel electrophoresis), 12.5 % gel composition (500 mL for 6 gels): 200 mL 30 % (w/v) acrylamide/bisacrylamide, 186.5 mL 1 M Tris–HCl pH 8.8, and 5 mL 10 % SDS; make up to 500 mL with distilled water. Sonicate complete solution and add 2.5 mL 10 % ammonium persulphate (APS) and 0.25 mL *N,N,N,N′*-tetramethylethylenediamine (TEMED) before pouring the solution.
7. SDS electrophoresis running buffer: 25 mM Tris, 192 mM glycine and 0.2 % SDS. Store at room temperature.

2.6 Staining of Gels and Analysis of Gel Images

1. Fixation solution: absolute methanol (12 %) and acetic acid (7 %).
2. Silver Staining: PlusOne Silver Staining Kit, protein (GE Healthcare), containing sensitizing solution 30 % (v/v) EtOH, 0.2 % (w/v) $Na_2S_2O_3$, 6.8 % (w/v) CH_3COONa, silver solution (0.25 % $AgNO_3$), developing solution (2.5 % Na_2CO_3, 0.015 % (v/v) HCOH), and stopping solution (1.46 % (w/v) EDTA).

3. For visualization of CyDye fluorescently labeled proteins, Typhoon 9400 (GE Healthcare).
4. For image analysis, the DeCyder v7.0 software (GE Healthcare).

2.7 Spot Digestion

The following buffers should be freshly prepared, as they degrade very quickly:

1. Reduction buffer: 10 mM DTT (GE Healthcare) in 50 mM NH_4HCO_3 (99.5 % purity; Sigma Chemical).
2. Alkylation buffer: 55 mM iodoacetamide (Sigma Chemical) in 50 mM NH_4HCO_3.
3. Trypsin (sequencing grade; Promega) at a final concentration of 7.5 ng/μl in 50 mM NH_4HCO_3.
4. 0.2 % trifluoroacetic acid (TFA) (99.5 % purity, Sigma Chemical).

The following material is required:

1. Pierced V-bottom 96-well polypropylene microplates (Bruker Daltonik).
2. Proteineer DP protein digestion station (Bruker Daltonik).

2.8 Sample Preparation for MALDI-MS

1. Matrix solution: 0.2 g/l α-cyano-4-hydroxycinnamic acid (Bruker Daltonik) in 50 % aqueous acetonitrile and 0.2 % trifluoroacetic acid.
2. 600 μm AnchorChip prestructured MALDI probe (Bruker Daltonik).
3. Ultraflex MALDI-TOF/TOF mass spectrometer (Bruker Daltonik) with an automated analysis loop controlled by the FlexControl 2.2 software (Bruker Daltonik).

3 Methods

3.1 Patient Recruitment and Samples

Recruit thrombus samples from AAA patients who have undergone surgical repair. Written consent must be given by the patients.

3.2 AAA-Thrombus-Conditioned Media Isolation

1. Dissect the human AAA-thrombus samples under sterile conditions into luminal and abluminal parts, respectively, at the interface with circulating blood and with the remaining media (Fig. 1) (*see* **Note 1**).
2. Cut luminal and abluminal layers into small pieces (5 mm^3) and incubate them separately in RPMI 1640 medium containing antibiotics and an antimycotic (Gibco) for 24 h at 37 °C (6 mL/g of wet tissue) (*see* **Note 2**).

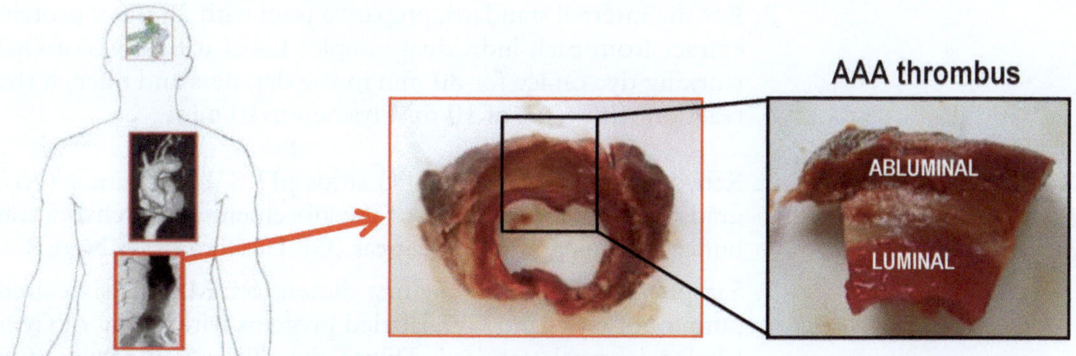

Fig. 1 Detailed inspection of an abdominal aortic aneurysm fragment, showing a section of the intraluminal thrombus. Abluminal and luminal layers are delimited by different cellular composition. Adapted from Fontaine V, Touat Z, Mtairag el M, Vranckx R, Louedec L, Houard X, Andreassian B, Sebbag U, Palombi T, Jacob MP, Meilhac O, Michel JB. Role of leukocyte elastase in preventing cellular re-colonization of the mural thrombus. Am J Pathol, 2004, 164:2077–2087

3. Centrifuge conditioned media (supernatant containing proteins released by the tissue sample) at 3,000 × g for 10 min at 20 °C to eliminate particulates.

3.3 Experimental Design and DIGE Protein Labeling

1. Isolate supernatants from luminal and abluminal layers of ILT human samples from six individual patients.
2. Precipitate proteins using the 2D clean-up kit and resuspend them in lysis buffer. Make sure that proteins are fully resuspended. Measure protein concentration by RC-DC protein assay (*see* **Note 3**).
3. Prepare stock CyDye reagents by adding 5 μL of dimethylformamide (DMF) to each of three CyDye vials. Spin and store at −80 °C until use (*see* **Note 4**).
4. Freshly prepare working CyDye solutions. For 6 gels (12 protein samples), a final volume of 6 μL per CyDye vial is required. For each CyDye use the following volumes:
   (a) Volume needed from stock solution: 6 μL/μmol (CyDye) × 0.4 μmol = 2.4 μL.
   (b) Volume needed of DMF: 6 − 2.4 μL = 3.6 μL.
   (c) Working solution preparation: 2.4 μL (Stock solution) + 3.6 μL (DMF) = 6 μL.
5. Spin-down samples and prepare individual samples with 50 μg of protein extract (*see* **Notes 5** and **6**).
6. Label individual samples with Cy3 or Cy5 dyes by adding 1 μL of working solution and spin-down. Incubate on ice for 30 min in the darkness (*see* **Note 7**).
7. Quench the reaction with 1 μL of 10 mM lysine for 10 min in the dark on ice.

8. For the internal standard, prepare a pool with 25 μg of protein extract from each individual sample. Label with 6 μL of Cy2 working dye on ice for 30 min in the darkness and quench the reaction with 6 μL of 10 mM lysine for 10 min.

3.4 Two-Dimensional Electrophoresis and Image Acquisition

1. Rehydrate six 24 cm-long IPG strips pH 3–11 nonlinear (NL) gradient with 450 μL each of the aforementioned rehydration buffer containing 97 mM DeStreak (GE Healthcare) (*see* **Note 8**).
2. Sample preparation for the first dimension: Mix the six-paired samples of Cy3- and Cy5-labeled proteins with 50 μg of Cy2-labeled internal standard. Dilute the mixture in rehydration buffer containing 50 mM DTT and resolve them by applying the sample mixtures via cup loading to the previously rehydrated IPG strips (*see* **Note 9**).
3. Isoelectric focusing (IEF): The isoelectric focusing can be carried out on an IPGphor II IEF system (GE Healthcare) up to a total of 42 kV h (*see* **Note 10**).
4. IPG strips equilibration: Equilibrate the strips in 10 mL of reducing SDS equilibration buffer during 15 min at room temperature with agitation. Then, eliminate the reducing solution and incubate with alkylating SDS equilibration buffer for 15 additional min (*see* **Notes 11** and **12**).
5. Second dimension: SDS-PAGE is carried out on 12 % polyacrylamide gels at 2 w/gel (for overnight running) or after sample entry in the running gel (40 min), increase to 20 w/gel (max. 100 w) for in-day gel running (*see* **Notes 13–15**).
6. Image acquisition: For each gel, acquire the Cy2-, Cy3-, and Cy5-labeled channel image at excitation/emission wavelength values of 488/520, 523/580, and 633/670 nm, respectively, using a Typhoon 9400 laser scanner (*see* **Note 16**).

3.5 DIGE Image Analysis

Analyze the images using the DeCyder v7.0 software for spot detection and quantification and inter-gel matching and statistics:

1. DIA (Differential In-gel Analysis): This is the DeCyder module for the initial spot detection, which compares all three fluorescence images obtained from each gel and performs co-detection, background removal, and normalization. The outcome is the volume of each spot detected in Cy3- and Cy5-labeled samples after normalization with respect to the corresponding Cy2 signal measured from the pooled internal standard sample. The DIA data sets from each individual gel were collectively analyzed using the biological variation analysis (BVA).
2. BVA: This is the second DeCyder module used for matching multiple 2D-DIGE gels, which provides statistical analysis of

protein abundance changes. Consider spots present in all of the 12 images (Cy2-, Cy3-, and Cy5-channel images for each gel) with statistical significance at 95 % confidence level for standardized average spot volume ratios over 1.5.

3. EDA (Extended Data Analysis): The third DeCyder module for advanced statistical analysis generates the final list of candidate spots, Principal Component Analyses (PCAs), and their hierarchical clustering, for the characterization and classification of biological samples based on protein expression data (*see* **Note 17**).

3.6 Gel Silver Staining

Stain the gels using PlusOne Silver Staining Kit from GE Healthcare with some modifications to the manufacturer's instructions (*see* **Note 18**). Briefly:

1. Incubate gels individually with Fixation solution at least for 1 h at 20 °C for protein fixing and gel cleaning (*see* **Note 19**).
2. Incubate gels individually with cold sensitizing solution for 20–30 min (*see* **Note 20**).
3. Wash with distilled water, three times 10 min each.
4. Incubate with 10 % silver solution (silver nitrate) for 20 min for silver impregnation.
5. Wash with distilled water, twice, 1 min each (*see* **Note 21**).
6. Add developing solution containing 0.4 % formaldehyde freshly supplemented for silver staining. Incubate gels until spots are visible (usually <5 min) with vigorous shaking.
7. Incubate gels with stopping solution for at least 15 min to stop silver reduction.
8. Wash three times, 15 min each, with distilled water and change the gel container.

3.7 In-Gel Tryptic Digestion

Digest samples automatically using a Proteineer DP protein digestion station following the protocol of Shevchenko et al. (5) with minor variations:

1. Spot picking: Match silver-stained gels against DIGE images and excise selected spots manually. Transfer them to pierced V-bottom 96-well microplates loaded with ultrapure water.
2. Submit gel plugs to reduction with 10 mM DTT (GE Healthcare) in 50 mM NH_4HCO_3 (99.5 % purity; Sigma).
3. Submit to alkylation with 55 mM iodoacetamide (Sigma Chemical) in 50 mM NH_4HCO_3.
4. Rinse gel pieces with 50 mM NH_4HCO_3 and acetonitrile (gradient grade; Merck) and dry them under a nitrogen stream.

5. Protein digestion: Add modified porcine trypsin at a final concentration of 7.5 ng/μl in 50 mM NH_4HCO_3 to the dried gel pieces and allow the digestion to proceed at 37 °C for 10 h.
6. Transfer the resulting digestion solutions by centrifugation to V-bottom 96-well polypropylene microplates, vacuum-dry them, and keep at 4 °C for later MS analysis.

3.8 MALDI Mass Spectrometry

1. Redissolve the dried samples in 10 μl of matrix solution.
2. Deposit 0.6 μl of this solution onto the 600 μm AnchorChip prestructured MALDI probe (6).
3. Allow the sample-matrix drops to dry at room temperature.
4. Analyze samples on an Ultraflex MALDI-TOF/TOF mass spectrometer (7) using an automated analysis loop controlled by the FlexControl 2.2 Software.
5. First step: Acquire MALDI-MS spectra by averaging 400 individual spectra in the positive ion reflector mode at 50 Hz laser frequency in the 800–3,500 *m/z* range. Perform internal calibration of MALDI-MS mass spectra using the two trypsin autolysis ions showing at 842.510 and 2,211.105 *m/z* (*see* **Note 22**).
6. Second step: Acquire MALDI-MS/MS spectra by averaging 800 individual spectra from precursor ions showing with intensity >1,500 in the above MALDI-MS spectra.
7. Perform an automated analysis of mass data using the flex Analysis 2.2 software to obtain peptide (MALDI-MS) and precursor (MALDI-MS/MS) *m/z* values.

3.9 MALDI-MS Database Searching

1. Combine MALDI-MS and MS/MS data through the BioTools 3.0 program to search a nonredundant protein database (NCBInr 20091015, ~10^7 entries, National Center for Biotechnology Information, Bethesda US), using the Mascot software v2.2 (Matrix Science, London, UK; http://www.matrixscience.com) (8).
2. Set other relevant search parameters as follows: enzyme, trypsin; fixed modifications, carbamidomethyl; allow up to 1 missed cleavage; peptide tolerance ±20 ppm; MS/MS tolerance ±0.5 Da.
3. Consider protein scores significant when greater than 82 ($p<0.05$) (*see* **Note 23**).

4 Notes

1. Tissue samples must be washed with saline buffer before their incubation in the culture medium to minimize blood remains.
2. Sterile material is required for tissue dissection.

3. Supernatants or conditioned media must be precipitated before IEF to remove salts and other ionic components that could interfere in both the CyDye labeling reaction and later in the isoelectrophoretic separation. To facilitate complete resuspension of protein pellets, shake vigorously for at least 1 h at room temperature.
4. After reconstitution, dye reagents (stocks solutions) are stable for three months at −70 °C/−80 °C, while working solutions are stable for just 1 week at −70 °C/−80 °C.
5. Check that pH of protein extracts is *ca.* 8.5 to ensure appropriate Fluor dye labeling. The pH must be checked on ice after protein solubilization and when necessary carefully modified by addition of small volumes of Tris–HCl (1 M; pH 9.0) or aqueous sodium hydroxide.
6. Under minimal labeling conditions, 400 pmol Fluor CyDye labels 50 μg of protein.
7. To ensure adequate and sustained protein labeling, not only the labeling reaction must be performed on ice and in the dark, but also the labeled samples must be strictly kept under these conditions.
8. It is recommended that the rehydration of IPG strips be completed overnight (or at least 10–12 h) at around 20 °C. The recommended volume must be used to ensure a homogeneous strip rehydration and conditioning.
9. For optimal labeling, the volumes must be adjusted in the 3–5 mg/ml range. Furthermore, it is advisable to use comparable loading volumes for all the strips (120 μL).
10. IEF conditions recommended for 24 cm, pH 3–11 IPG strips are as follows: Step 1, 300 V, 3 h, gradient; Step 2, 1,000 V, 5 h, gradient; Step 3, 8,000 V, 2 h, gradient; Step 4, 8,000 V, constant up to a total of 42 kV h. A final step of 500 V is also recommended to avoid protein diffusion after focusing.
11. In the alkylating SDS equilibration buffer, 3.5–4 % iodoacetamide (IAA) is recommended. During the second equilibration step, IAA must be in excess with respect to DTT to guarantee that Cys residues will remain protected by alkylation. This excess IAA not only avoids reformation of sulfur–sulfur bonds but also prevents DTT point streaking in gels.
12. If the second dimension is not run immediately, IPG strips can be stored at −20 °C for a few weeks. Do not store the strips once they have been equilibrated.
13. Gels must be polymerized overnight at *ca.* 20 °C to reduce acrylamide-modified artifacts. No stacking gel is necessary as IPG strips works as a stacking gel.

14. 0.2 % SDS-running buffer is recommended instead of the usual 0.1 % SDS buffer.
15. The second dimension can be run at 2 w/gel and higher potential values, up to a maximum of 17 w/gel. Control temperature (e.g., 20 °C) for optimal protein separation.
16. Gel pre-scanning is advised before acquiring the final image for DeCyder analysis. Scanner photomultiplier values must be adjusted in the three channels (Cy2, Cy3, and Cy5) to avoid saturation of the spot signals. It is also recommended to use comparable maximum values of pixel intensities for the three images.
17. Results derived from EDA module show data quality, in order to delete outlier values if needed.
18. To ensure compatibility with later MS analyses, glutardialdehyde and formaldehyde must not be added to sensitizing and silver solutions, respectively. To compensate for the slightly decreased sensitivity of the as-prepared silver nitrate solution, the developing solution is supplemented with twice the usual concentration of formaldehyde.
19. As a gel fixing solution, methanol is preferred rather than ethanol, as its lower reactivity with acetic acid reduces unwanted protein esterification.
20. The use of a cold sensitizing solution is recommended to minimize gel background.
21. The gel must be completely submerged into water during the washing steps. Make sure that the final washing steps before the developing process last strictly 1 min each to avoid removal of silver ions.
22. For MALDI-MS/MS, the calibrations must be performed with fragment ion spectra obtained for the proton adducts of a peptide mixture covering the 800–3,200 m/z region.
23. The absolute value of the score for statistical significance varies from experiment to experiment, which depends on the sequence database and on the used searching parameters.

Acknowledgments

This chapter has been supported by the EC, FAD project (FP-7, HEALTH F2-2008-200647), the Spanish MICIN (SAF2010/21852), Ministerio de Sanidad y Consumo, Instituto de Salud Carlos III, Redes RECAVA (RD06/0014/0035), EUS2008-03565, and Fundacion Pro CNIC.

References

1. Blanco-Colio LM et al (2009) Vascular proteomics, a translational approach: from traditional to novel proteomic techniques. Expert Rev Proteomics 6:461–464
2. Nordon I et al (2009) The role of proteomic research in vascular disease. J Vasc Surg 49:1602–1612
3. Martinez-Pinna R et al (2008) Proteomics in atherosclerosis. Curr Atheroscler Rep 10:209–215
4. Martinez-Pinna R et al (2011) Identification of peroxiredoxin-1 as a novel biomarker of abdominal aortic aneurysm. Arterioscler Thromb Vasc Biol 31:935–943
5. Shevchenko A et al (2006) In-gel digestion for mass spectrometric characterization of proteins and proteomes. Nat Protoc 1:2856–2860
6. Schuerenberg M et al (2000) Prestructured MALDI-MS sample supports. Anal Chem 72:3436–3442
7. Suckau D et al (2003) A novel MALDI LIFT-TOF/TOF mass spectrometer for proteomics. Anal Bioanal Chem 376:952–965
8. Perkins DN et al (1999) Probability-based protein identification by searching sequence databases using mass spectrometry data. Electrophoresis 20:3551–3567

Chapter 9

Metabolites Secreted by Human Atherothrombotic Aneurysm

Michal Ciborowski and Coral Barbas

Abstract One of the possibilities to study abdominal aortic aneurysm (AAA) biomarkers is to use aneurysm biopsies of dilated arteries or intraluminal thrombus (ILT) for explant cultures. Those metabolites secreted into the culture medium are concentrated in comparison to plasma and free from other interferences that can appear in it, allowing an easier harvesting. Liquid chromatography coupled to quadrupole time-of-flight mass spectrometry detector (LC-QTOF-MS) with an electrospray (ESI) interface has a broad applicability to detect metabolites of all classes with high sensitivity and mass accuracy. Therefore, an LC-QTOF-MS-based metabolomics approach was chosen to select and identify metabolites that could be useful as diagnostic/prognostic markers of AAA. This chapter describes the methodology for the differential analysis of AAA metabolites in secretomes. It gives experimental details on basic steps such as sample preparation, protein precipitation, and analysis of the samples by LC-MS. A description for quality control of the methodology, raw data processing, as well as selection and identification of differentiating metabolites are also presented in this chapter.

Key words Abdominal aortic aneurysm, Secretome, Metabolic fingerprinting, LC-QTOF-MS, Multivariate analysis, Metabolites identification

1 Introduction

Abdominal aortic aneurysm (AAA) is a permanent and localized aortic dilation, defined as aortic diameter ≥3 cm (1). It is an asymptomatic but potentially fatal condition because progressive enlargement of the abdominal aorta is spontaneously evolving towards rupture. Additionally, inside a dilated aorta an intraluminal thrombus (ILT) could be formed. ILT may be involved in the evolution and possible rupture of AAA (2). Although it is hypothesized that the incidence of AAAs has begun to fall in Australia and that the admission rate for aneurysm repair was observed to decline in USA (3), ruptured AAAs are still responsible for at least 15,000 deaths in the USA each year (4). Efforts to limit the mortality rate from AAA rupture depend on early detection and elective AAA repair. However, most AAAs show discontinuous growth patterns and

Fernando Vivanco (ed.), *Vascular Proteomics: Methods and Protocols*, Methods in Molecular Biology, vol. 1000, DOI 10.1007/978-1-62703-405-0_9,

alternate periods of stability and nongrowth with periods of acute expansion and occasionally ruptures (5). For that reason, biomarkers of growth and rupture could give us a more nuanced indication for surgery and could also afford novel pathogenic pathways, thus opening possibilities for pharmacological inhibition of growth. As direct analysis of blood/plasma, a common and noninvasive patient screening method can be difficult for candidate biomarker identification, alternative/complementary approaches are required, and one of them is the analysis of secretomes in cell culture media in vitro. As the compounds secreted by cells are most likely secreted into blood/plasma, the identification and preselection of candidate biomarkers from cell secretomes with subsequent validation of their presence in serum/plasma is a promising approach. Primarily, this strategy was proposed for differential analysis of proteins released by normal and pathological arteries as well as atherosclerotic plaques in culture. In this methodology, surgical pieces of arterial wall and atherosclerotic plaques were cultured in protein-free RPMI culture medium with additives. After incubation, proteins secreted by the wall or plaque to the culture medium were separated by 2-D gel electrophoresis and identified by MALDI-TOF mass spectrometry. This proteomic strategy allows identification of proteins secreted from atherosclerotic plaques that can be used as biomarkers of the stage of the pathology (6).

We have used a similar strategy to study the rate of exchange of metabolites between arteries or ILT obtained from AAA patients and the external medium. AAA thrombus samples, as well as aneurysmal and healthy walls, were cut into small pieces and separately incubated in RPMI 1640 medium. To analyze secreted metabolites, we used an LC-QTOF-MS-based metabolomics approach, in particular metabolic fingerprinting which looks into a total profile, or fingerprint, as a unique pattern characterizing metabolism (7). In this way, we tried to find those molecules that would have a higher probability of later being found in plasma and could start signaling processes or be useful as diagnostic/prognostic markers (8).

This chapter provides experimental details on basic steps such as sample preparation, protein precipitation, and analysis of the samples by LC-MS. A description for quality control of the methodology, raw data processing, as well as selection and identification of differentiating metabolites are also presented.

2 Materials

2.1 Chemicals

1. RPMI 1640 medium containing antibiotics and antimycotic (Gibco®) or similar (*see* **Note 1**).
2. Standard compounds (Sigma, St. Louis, MO, USA): oleamide, palmitoylcarnitine, docosahexaenoic acid, 5-oxo-proline, hippuric acid, *O*-phosphotreonine, alpha-tocopherol, octylamine.

3. LC-MS-grade acetonitrile.
4. Formic acid for mass spectrometry ≥99.5 % (Fluka, Steinheim, Germany) or similar.
5. Ultrapure water can be produced by a Milli-Q Reagent Water System (Millipore, MA, USA) or similar.

2.2 Samples

AAA thrombus samples collected during surgical repair were dissected into luminal and abluminal parts (respectively at the interface with circulating blood and with the remaining media) (*see* **Note 2**). Control aortas were sampled from dead organ donors (*see* **Note 3**).

2.3 Equipment

1. Microcentrifuge Eppendorf 5415R (Eppendorf AG, Hamburg, Germany).
2. Vortex mixer.

2.4 LC-MS Equipment

1. 1200 series HPLC (Agilent Technologies, St. Clara, CA, USA).
2. Thermostated autosampler 1290 series (Agilent Technologies, St. Clara, CA, USA).
3. LC column: Discovery HS C18 15 cm × 2.1 mm, 3 μm (Supelco) with a guard column: Discovery HS C18 2 cm × 2.1 mm, 3 μm (Supelco).
4. Accurate mass quadrupole time-of-flight MS detector (QTOF) with electrospray ionization (ESI) technique, model 6520 (Agilent Technologies, St. Clara, CA, USA).

2.5 Software

1. Mass Hunter Qualitative Analysis Software (Agilent Technologies, St. Clara, CA, USA). This software is used for reading, displaying, and performing several calculations with obtained chromatograms.
2. Mass Profiler Professional 2.2 (Agilent Technologies, St. Clara, CA, USA). Among other possibilities, this software is used for alignment of chromatograms and data filtering.
3. SIMCA P+ version 12 (Umetrics, Umeå, Sweden). This software is used for multivariate analysis of data.
4. MS Excel (Microsoft®).

2.6 Others

1. 0.22 μm nylon syringe filters (National Scientific, Rockwood, TN, USA).
2. 1 mL syringes Soft-Ject (Henke sass wolf, Tultlingen, Germany).
3. 2 mL GC-HPLC vials with 200 μL conical narrow opening inserts.

3 Methods

3.1 Secretomes Preparation (See Note 4)

1. Cut dissected luminal and abluminal layers of AAA thrombus, as well as aneurysmal and healthy walls, into small pieces (5 mm^2).
2. Weight cut pieces.
3. Incubate obtained samples separately in RPMI 1640 medium for 24 h at 37 °C, add 6 mL of medium per 1 g of wet tissue.
4. Centrifuge samples at 3,000 × *g* for 10 min at 20 °C to obtain conditioned medium (supernatant containing proteins and metabolites released by the tissue sample).
5. From each sample take 250 μL of the supernatant, centrifuge at 14,000 × *g*, and determine the protein concentration (e.g., by Bradford's (9) method). Remaining supernatant can be frozen and stored at −80 °C in aliquots prior to analysis.

3.2 Sample Preparation for LC-MS Fingerprinting

Currently, there are several analytical techniques used in metabolomics studies like nuclear magnetic resonance (NMR), high-performance liquid chromatography/mass spectrometry (LC/MS), gas chromatography/mass spectrometry (GC/MS), and capillary electrophoresis/mass spectrometry (CE/MS). Each of them has its own drawbacks and advantages (10). Each technique requires appropriate sample and data treatment. In this case, we used LC-MS-based metabolomics to study AAA biomarkers. Application of LC-MS with an ESI interface and QTOF as detector has a broad applicability to detect metabolites of all classes with high sensitivity and mass accuracy, and this makes it the most versatile choice for analysis and studies of new biomarkers (8, 11).

1. Protein precipitation and metabolite extraction:
   (a) Prior to further manipulation, secretome samples must be thawed and vortex-mixed (*see* **Note 5**).
   (b) From each sample, take equal volumes and pool them in one Eppendorf (*see* **Notes 6** and 7). This will be the quality control sample (QC).
   (c) On ice, pipette 50 μL of each sample (including QC) to an Eppendorf, vortex-mix it with 200 μL of cold (−20 °C) acetonitrile for 1 min and let stand for 5 min (*see* **Note 8**).
   (d) Centrifuge the sample at 16,000 × *g* for 10 min at 4 °C.
   (e) Filter supernatant through 0.22 μm syringe nylon filter directly to the LC vial equipped with a 200 μL insert (*see* **Note 9**).

3.3 Metabolite Fingerprinting with LC-MS

Metabolites expected to be present in the samples should be similar to those found in plasma; therefore we used chromatography and mass spectrometry conditions previously optimized and utilized for metabolic fingerprinting of plasma samples (11).

1. Inject 10 μL of the treated secretome sample into the LC-MS system. Maintain autosampler at 4 °C.
2. Maintain column oven temperature at 40 °C.
3. Set the capillary voltage to 3,000 V and the nebulizer gas flow rate to 10.5 L/min.
4. Operate mass spectrometer in scan mode from 50 to 1,000 *m/z* with a scan rate of 1.02 scan per second.
5. Operate the HPLC system at the flow rate of 0.6 mL/min with solvent A, water with 0.1 % formic acid, and solvent B, acetonitrile with 0.1 % formic acid. Start the gradient from 25 % B to 95 % B in 35 min and return to starting conditions in 1 min, keep the re-equilibration at 25 % B for 9 min.
6. Analyze samples in one randomized run using also QC samples (*see* **Note 6**).

3.4 Data Analysis

There are several licensed and open-access softwares for LC-MS-based metabolomics data pre-processing and analysis. In our laboratory, data were treated as described:

1. Quality control of the methodology
   (a) Check the quality of Total Ion Chromatogram (TIC) obtained for each sample and QCs.
   (b) Open pump pressure curves for every individual analysis (*see* **Note 10**).
   (c) In Mass Hunter Qualitative Analysis Software, perform Molecular Feature Extraction for each sample and QCs (*see* **Note 11**).
   (d) For each sample, export a list of created compounds as a cef file (*see* **Note 12**).
   (e) Open all cef files in Mass Profiler Professional Software and perform samples alignment.
   (f) After alignment, export a txt file with the matrix of data with one column per sample (include all the secretome samples and QCs) and one row per any individual metabolite detected in all samples (*see* **Note 13**).
   (g) Open exported file in SIMCA and build principal components analysis (PCA), principal least squares discriminant analysis (PLS-DA), or orthogonal projection to latent structure-discriminant analysis (OPLS-DA) model based on secretome samples. Use this model to predict the position of QC samples (*see* **Note 14**).
2. Selection of statistically significant variables
   (a) Open secretomes cef files in Mass Profiler Professional Software and perform again sample alignment, now without QC samples.

(b) Once your samples are aligned, load the data matrix to excel and perform normalization of the data by dividing the intensities of metabolites by the protein concentration in each sample (*see* **Note 15**).

(c) Load the normalized matrix to appropriate software in order to perform multivariate data analysis (PCA, PLS-DA, or OPLS-DA).

(d) Once sample cluster according to the classes and chemometric models are adequately validated, identify variables responsible for the samples classification. Statistically significant variables can also be selected by univariate statistical analysis (e.g., *t* test).

3.5 Metabolites Identification

1. Introduce accurate mass of statistically significant variables into accurate mass databases (DBs) in order to find metabolites having similar mass (*see* **Note 16**).
2. Check the isotopic pattern of identified metabolite with isotopic pattern of your metabolite (*see* **Note 17**).
3. In order to confirm the identity of metabolites resulting from DBs searching, analyze the standard of this metabolite (if available) or perform LC-MS/MS analysis with the same equipment and identical chromatographic conditions to the primary analysis (*see* **Notes 18** and **19**).

The final result of the metabolomics experiment is a list of metabolites significantly increasing or decreasing between the groups of samples under investigation. Significantly changing metabolites identified in secretomes from human atherothrombotic aneurysm are listed in Table 1. However, the most important value of a metabolomics study is the biological interpretation of the meaning of observed changes. Therefore, much effort should be devoted for proper interpretation of obtained results.

4 Notes

1. Try to use for the whole experiment medium from the same bottle or at least from the same batch.
2. Both parts are visibly different in color (luminal is red while abluminal is dark brown).
3. These control aortic samples should be macroscopically normal, devoid of early atheromatous lesions.
4. This step must be performed in sterile conditions by use of sterile (autoclaved) materials (pipettes, tips, Eppendorfs).
5. Samples thawed more than twice are not recommended.

Table 1
Metabolites identified in secretomes from human atherothrombotic aneurysm

Compound	*RT* (min)	Measured mass (Da)	Formula	Identification
Palmiticamide	26.2	255.254	$C_{16}H_{33}NO$	256.262, 116.107, 102.092, 88.075, 57.069
Stearamide	31.0	283.2862	$C_{18}H_{37}NO$	284.295, 116.107, 102.090, 88.076, 57.070
Oleamide	27.0	281.2698	$C_{18}H_{35}NO$	Standard
Anandamide	25.1	351.3127	$C_{22}H_{41}NO_2$	352.307, 335.282, 103.073, 62.06
Palmitoylcarnitine	17.8	399.3316	$C_{23}H_{45}NO_4$	Standard
Stearoylcarnitine	20.9	427.3631	$C_{25}H_{49}NO_4$	428.3734, 311.289, 73.024
Palmitic acid	27.5	256.2387	$C_{16}H_{32}O_2$	257.247, 117.088, 103.074, 89.059, 71.086, 57.07
Stearic acid	33.8	284.2692	$C_{18}H_{36}O_2$	285.294, 240.233, 116.054, 71.087, 57.07
Lyso PC	19.6	495.3292	$C_{24}H_{50}NO_7P$	496.341, 184.069, 104.103, 86.095
Lyso PC	24.2	523.3598	$C_{26}H_{54}NO_7P$	524.368, 184.072, 104.107, 86.096
GPCho	34.4	636.4571	$C_{33}H_{67}NO_8P$	637.465, 311.294, 283.259, 184.07
GPCho	34.9	548.4051	$C_{29}H_{59}NO_6P$	549.413, 311.293, 184.072
Docosahexaenoic acid	26.5	328.2395	$C_{22}H_{32}O_2$	Standard
5-oxo-proline	0.7	129.0421	$C_5H_7NO_3$	Standard
Glutamylhydroxyproline	6.6	260.1009	$C_{10}H_{16}N_2O_6$	261.106, 217.188, 173.085
Guanidinosuccinic acid	0.7	175.0572	$C_5H_9N_3O_4$	176.065, 130.085, 118.121, 100.096, 70.056, 60.056
Hippuric acid	0.9	179.0573	$C_9H_9NO_3$	Standard
O-phosphothreonine	0.6	199.0318	$C_4H_{10}NO_6P$	Standard
Hydroxyethyl-methylthiazole	0.6	143.0396	C_6H_9NOS	144.045, 126.034,118.064, 113.028, 82.942
Alpha-tocopherol	29.9	430.373	$C_{29}H_{50}O_2$	Standard
Pyridoxamine 5 -phosphate	0.7	248.0518	$C_8H_{13}N_2O_5P$	249.059, 191.038, 135.003
Hydroxy-oxo-cholanoic acid	33.8	390.2747	$C_{24}H_{38}O_4$	391.284, 167.032, 149.02, 71.085, 57.07
Octylamine	0.8	129.1515	$C_8H_{19}N$	Standard

6. In LC-MS-based metabolomics, a method for monitoring the quality of the results and the overall analytical procedure is required (12). With that aim, quality control (QC) samples must be prepared. It is recommended that QC samples are a mixture of all samples analyzed in the experiment. Each QC sample is prepared independently following the same procedure as for the rest of samples. Those samples (each QC sample should be injected only once) are analyzed at the beginning, at the end, and throughout the run in order to provide a measurement not only of the system's stability and performance but also of the reproducibility of the sample treatment procedure.
7. Plan how many QC samples are needed in your experiment and pool adequate volume of secretome samples.
8. The main components of culture medium are inorganic salts, which disturb ionization in ESI technique. Salts aggregating in ion source inhibit the formation of useful analyte ions. Therefore, for simultaneous precipitation proteins and salts, acetonitrile is strongly recommended.
9. Other filters compatible with acetonitrile could also be used. However, you should always check the profile of filtered solvent (blank) obtained with your method.
10. Pressure curves for each analysis should be overlapping.
11. Molecular Feature Extraction (MFE) is a tool, which cleans a data file of extraneous background noise and unrelated ions, and creates a list of all possible compounds in chromatogram represented by the full TOF mass spectral data and retention time.
12. Cef is the type of the file, which can be opened in Mass Profiler Professional Software.
13. Each metabolite is represented by accurate mass and retention time.
14. If sample preparation procedure and condition of equipment during analysis were proper, QCs should cluster in the middle of the plot between your samples (Fig. 1).
15. As we observed variations in protein concentration between the cultures of thrombi (6.5 ± 2.5 μg/μL) and arteries (3.4 ± 1.2 μg/μL), data were normalized by dividing intensities of metabolites by the concentration of the protein in each sample. Other parameters to normalize data, like total signal or total useful signal (13), are also worth checking.
16. There are several free of charge, accurate mass DBs available on the internet, like: METLIN (http://metlin.scripps.edu), KEGG (http://www.genome.jp/kegg), LIPIDMAPS (http://www.lipidmaps.org/), and HMDB (http://hmdb.ca).

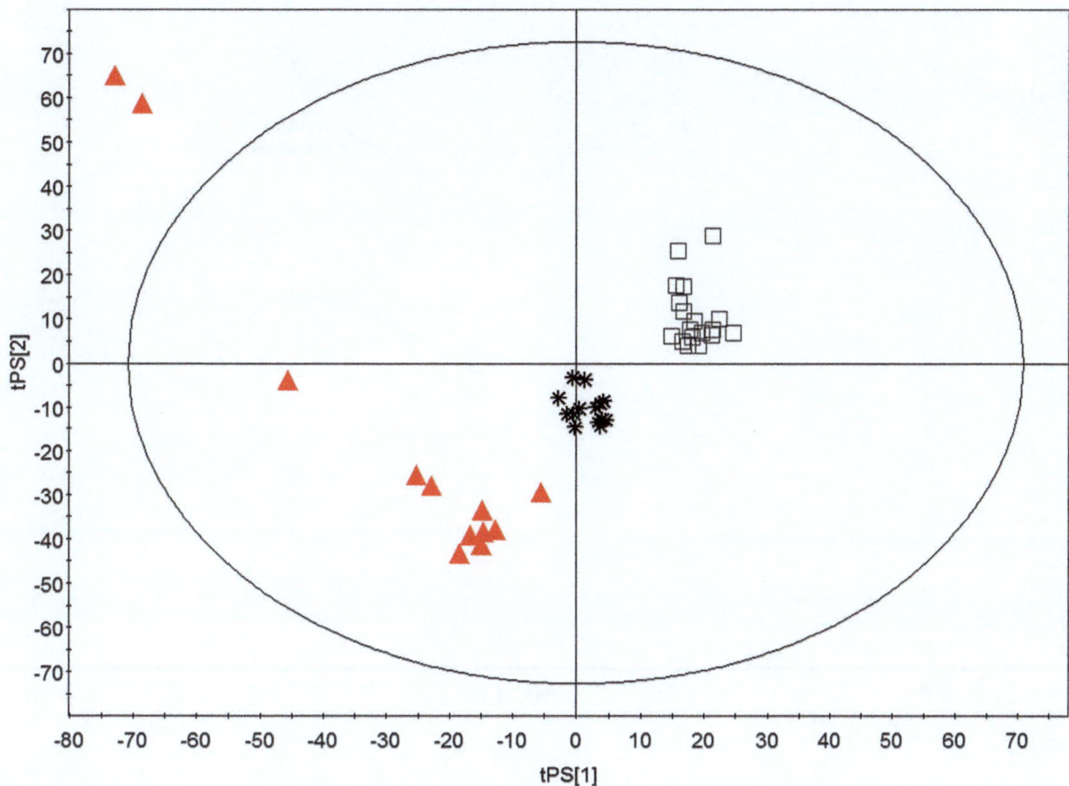

Fig. 1 Prediction for QC samples by PLS-DA model built for two groups of secretome samples. A PLS-DA model was built for two groups (*red triangle*—secretomes from arteries, *empty square*—secretomes from thrombi) taking all variables generated as molecular features in the mass spectrum. Variables were not normalized, but scaled by applying common logarithm to the abundances in order to approximate a normal distribution of the data. The parameters for the model are $R^2 = 0.973$, $Q^2 = 0.532$

17. An isotopic pattern is characteristic for the formula of compounds, so a theoretical isotopic pattern obtained for metabolite found in DBs should match this in a real sample. You can use open-source mass spectrometry tool like mMass (14) to generate theoretical isotopic patterns.
18. Add the standard to the real sample (spiking) to eliminate matrix effects. Analyze this sample before and after addition of the standard. Generate extracted ion chromatogram (EIC) for this mass in both samples. EICs for both samples should be overlapping in retention time, but the peak in the sample with the addition of the standard should be higher (Fig. 2, panel A).
19. Compare the obtained fragmentation pattern in MS/MS spectrum with the MS/MS spectrum for the standard (Fig. 2, panel B). MS/MS fragmentation patterns are available for some metabolites in mentioned DBs or scientific articles. Be aware of

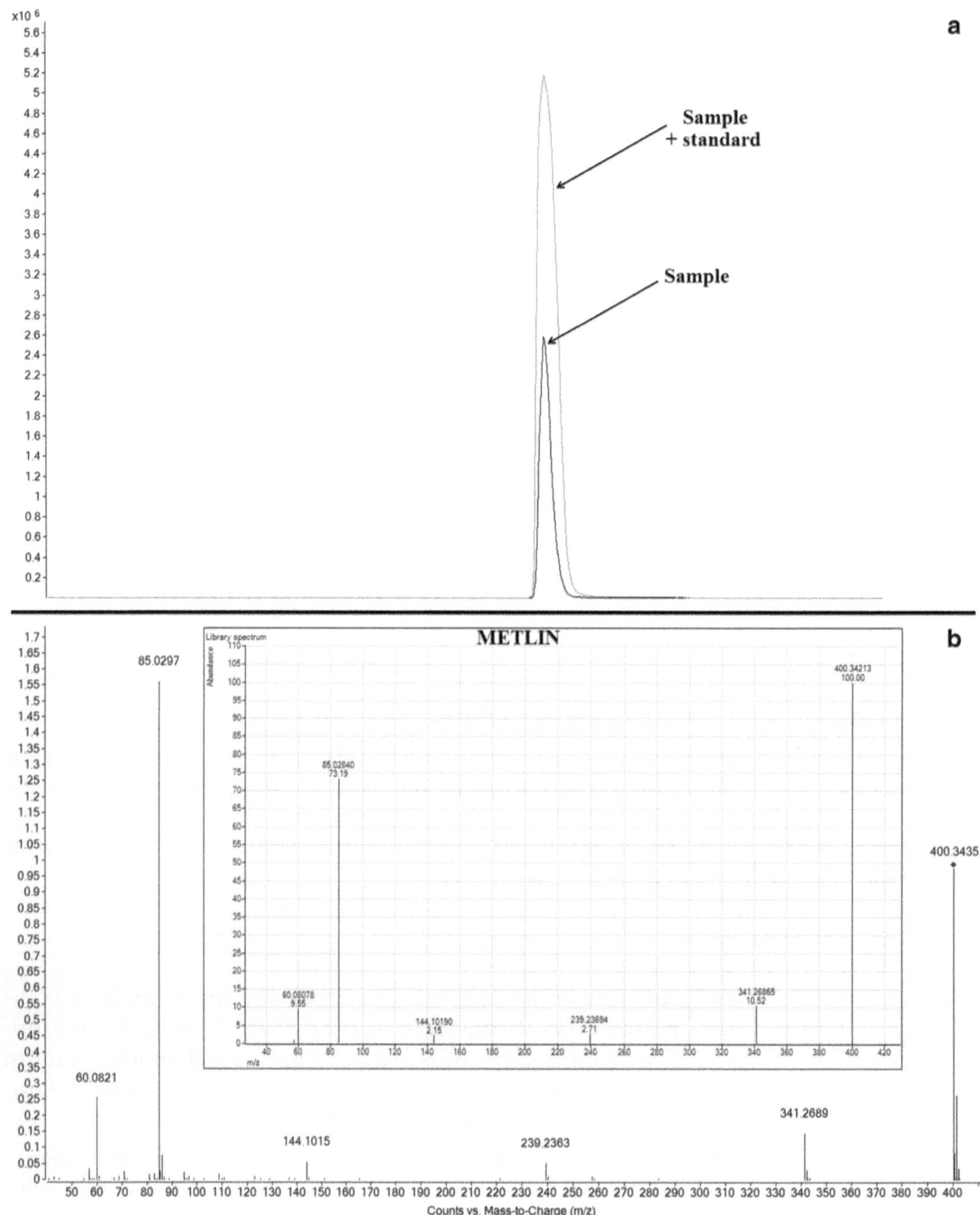

Fig. 2 Strategy for confirmation of metabolites identified in databases based on accurate mass. *Panel A* shows EIC for palmitoylcarnitine in secretome sample before and after addition of the standard of this metabolite. *Panel B* shows MSMS spectras of palmitoylcarnitine. One obtained by analysis of secretome sample and other found in METLIN database

the type of instrument and ionization technique that was used to obtain spectra presented in DBs. Different ionization techniques and MS detectors may result in different fragmentation patterns.

References

1. Miyake T, Morishita R (2009) Pharmacological treatment of abdominal aortic aneurysm. Cardiovasc Res 83:436–443
2. Touat Z, Ollivier V, Dai J, Huisse M, Bezeaud A, Sebbag U, Palombi T, Rossignol P, Meilhac O, Guillin M, Michel J (2006) Renewal of mural thrombus releases plasma markers and is involved in aortic abdominal aneurysm evolution. Am J Pathol 168:1022–1030
3. Levin DC, Rao VM, Parker L, Frangos AJ, Sunshine JH (2009) Endovascular repair vs open surgical repair of abdominal aortic aneurysms: comparative utilization trends from 2001 to 2006. J Am Coll Radiol 6:506–509
4. Klink A, Hyafil F, Rudd J, Faries P, Fuster V, Mallat Z, Meilhac O, Mulder WJ, Michel JB, Ramirez F, Storm G, Thompson R, Turnbull IC, Egido J, Martín-Ventura JL, Zaragoza C, Letourneur D, Fayad ZA (2011) Diagnostic and therapeutic strategies for small abdominal aortic aneurysms. Nat Rev Cardiol 8:338–347
5. Vega de Céniga M, Esteban M, Quintana J, Barba A, Estallo L, de la Fuente N, Viviens B, Martin-Ventura J (2009) Search for serum biomarkers associated with abdominal aortic aneurysm growth—a pilot study. Eur J Vasc Endovasc Surg 37:297–299
6. Duran M, Mas S, Martin-Ventura J, Meilhac O, Michel J, Gallego-Delgado J, Lázaro A, Tuñon J, Egido J, Vivanco F (2003) Proteomic analysis of human vessels: application to atherosclerotic plaques. Proteomics 3:973–978
7. Shulaev V (2006) Metabolomics technology and bioinformatics. Brief Bioinform 7: 128–139
8. Ciborowski M, Martin-Ventura JL, Meilhac O, Michel JB, Ruperez FJ, Tuñon J, Egido J, Barbas C (2011) Metabolites secreted by human atherothrombotic aneurysms revealed through a metabolomic approach. J Proteome Res 10:1374–1382
9. Bradford MM (1976) A rapid and sensitive method for the quantitation of microgram quantities of protein utilizing the principle of protein-dye binding. Anal Biochem 72:248–254
10. Lindon JC, Holmes E, Nicholson JK (2006) Metabonomics techniques and applications to pharmaceutical research & development. Pharm Res 23:1075–1088
11. Ciborowski M, Javier Rupérez F, Martínez-Alcázar M, Angulo S, Radziwon P, Olszanski R, Kloczko J, Barbas C (2010) Metabolomic approach with LC-MS reveals significant effect of pressure on diver's plasma. J Proteome Res 9:4131–4137
12. Gika H, Macpherson E, Theodoridis G, Wilson I (2008) Evaluation of the repeatability of ultra-performance liquid chromatography-TOF-MS for global metabolic profiling of human urine samples. J Chromatogr B Analyt Technol Biomed Life Sci 871: 299–305
13. Godzien J, Ciborowski M, Angulo S, Ruperez FJ, Martínez MP, Señorans FJ, Cifuentes A, Ibañez E, Barbas C (2011) Metabolomic approach with LC-QTOF to study the effect of a nutraceutical treatment on urine of diabetic rats. J Proteome Res 10:837–844
14. Strohalm M, Kavan D, Novák P, Volný M, Havlíček V (2010) mMass 3: a cross-platform software environment for precise analysis of mass spectrometric data. Anal Chem 82: 4648–4651

Chapter 10

Quantitative Analysis of Apolipoproteins in Human HDL by Top-Down Differential Mass Spectrometry

Matthew T. Mazur and Helene L. Cardasis

Abstract

The field of quantitative, label-free proteomics has evolved significantly over time, with most experiments performed "bottom-up" using proteolyzed protein mixtures. In these experiments, statistically significant peptide abundance differences between two or more experimental conditions are determined, and their corresponding proteins later identified. Recently, the rationale for extending this experimental design to mixtures of intact proteins has become clear, as analysis at the protein level allows for the independent detection of each protein form present, including those modified posttranslationally. This provides a level of specificity lost in bottom-up experiments. As such, the application of label-free top-down differential mass spectrometry has provided a means for understanding the subtle protein changes that define a particular phenotype. Described here is an approach for the top-down label-free quantitative analysis of the proteins which constitute human high-density lipoprotein particles. The methodology is conceptually very straightforward; however, it does require a level of rigor and consistency typically not addressed by more conventional proteomics experiments.

Key words Quantitative, Label-free proteomics, Top-down proteomics, High-performance mass spectrometry, Liquid chromatography-mass spectrometry (LC-MS), Differential mass spectrometry (dMS), High-density lipoprotein (HDL)

1 Introduction

Over the last few decades, the field of mass spectrometry-based proteomics has evolved from the large-scale identification of proteins within a given organism or system (1–3) to the relative quantitation of these same proteins under different experimental conditions or physiological states, e.g., normal vs. diseased. This quantitative approach aims to determine those proteins that play a functional role in the biological process of interest. Numerous technical approaches towards protein relative quantitation have emerged, including isobaric tagging for relative and absolute quantitation (iTRAQ) (4, 5), stable-isotope labeling by amino acid in cell culture (SILAC) (6, 7), and quantitative, label-free proteomics (8–13).

Fernando Vivanco (ed.), *Vascular Proteomics: Methods and Protocols*, Methods in Molecular Biology, vol. 1000,
DOI 10.1007/978-1-62703-405-0_10,

These techniques each allow for the relative quantitation of detectable proteins between two or more conditions; however, most require the proteolytic digestion of protein mixtures prior to liquid chromatography-mass spectrometry (LC-MS) analysis (i.e., bottom-up analysis). Although "proteolysis" allows LC-MS quantitative sampling of a very large number of unique precursor proteins, the intact protein information, including posttranslational modification and splice variants, is often lost. Conversely, top-down approaches measure proteins at the intact level, enabling the detection of protein isoforms and/or posttranslationally modified forms as independent species. For this reason, quantitative data generated by a top-down approach can be directly interpreted, resulting in accurate, relevant biological information (14). The ability to adopt a top-down experimental paradigm gives label-free proteomics, or differential mass spectrometry (dMS), a clear advantage for the type of work described here.

Most of the same advantages that exist for quantitative, label-free proteomics approaches used in bottom-up experiments remain when experiments are extended to intact protein mixtures. dMS analysis allows for the unbiased quantitative analysis of all ions detected within full-scan spectra of acquired LC-MS data and are easily amenable to multivariable experiments (10, 12), provided that a few key conditions are met. First, significant attention must be paid to the reproducibility of sample preparation techniques, as even slight deviations in sample preparation between samples can produce erroneous results. Next, chromatography must be optimized primarily for reproducibility. Chromatographic reproducibility and robustness is critical for proper feature assignment (a feature is defined by an accurate m/z and retention time), grouping, alignment, and quantitation by the applied algorithm (e.g., Elucidator PeakTeller™). As such, chromatographic resolution is often sacrificed at the expense of chromatographic reproducibility and optimal algorithm performance. Finally, because dMS approaches do not pool or chemically label samples prior to analysis, studies must be powered by a sufficient number of biological and technical sample replicates. As statistical analysis (e.g., t-test) is performed to determine those "features" of significance (Fig. 1), the greater the number of samples analyzed (i.e., degrees of freedom) allows more reliable statistical partitioning between the conditions under investigation.

The quantitative measurement of intact protein mixtures by top-down LC-MS requires some modification to procedures typical of corresponding bottom-up experiments, including the use of alternative sample preparation/introduction techniques, higher resolving power requirements for ion detection, and alternative technology and software for intact protein sequencing, identification, and characterization. Sample preparation for these experiments must be significantly altered from their counterpart

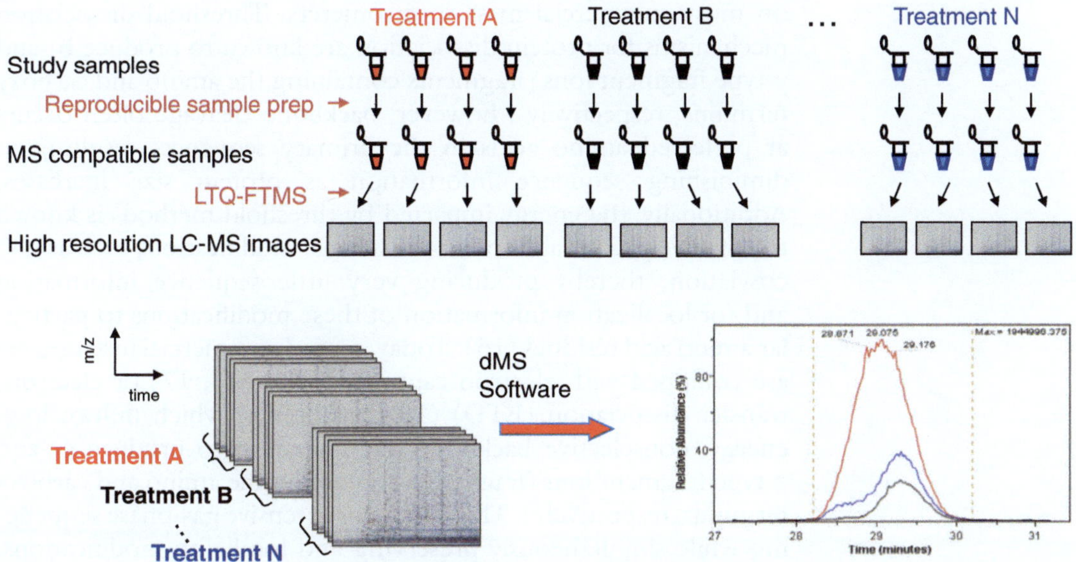

Fig. 1 The general experimental schematic of a differential mass spectrometry analysis. Replicate samples from multiple conditions (i.e., treatments A–N) are prepared for LC-MS analysis identically and reproducibly. Mass spectrometry compatible samples are then analyzed by LC-MS in a single sequence run block, and LC-MS data "images" are collected and processed by quantitative, label-free software to detect and extract features (i.e., accurate *m/z* and retention time). Statistically significant features are determined and later targeted for MS/MS identification

bottom-up strategies. While the removal of the enzymatic digestion step for top-down experiments simplifies the workflow, whole-protein liquid-phase concentration and clean-up methods must be considered and tested to ensure consistent sample processing. Further, although front-end liquid chromatography separation is intuitively analogous to bottom-up experiments, conditions must be modified to better accommodate larger protein analytes, which tend to retain more strongly to reverse-phase stationary phases than their peptide analogues. Liquid chromatography separation conditions are also intentionally simplified by use of linear gradients and tailored stationary phases (e.g., perfusion chromatography stationary phases) in order to best preserve the chromatographic reproducibility. Additionally, high-resolution detection plays a significant role in top-down quantitative profiling experiments, as analytes here will be relatively large and highly charged. Instruments such as Thermo's LTQ-FT Ultra provide highly sensitive detection of analytes at instrument resolving powers required for isotopic resolution of intact proteins and critical for the dMS analysis.

Unlike MS/MS dissociation of relatively low molecular weight peptide precursors, sequence identification and characterization of larger, highly charged protein ions can be challenging using conventional threshold dissociation mechanisms (i.e., collisionally induced dissociation or infrared multiphoton dissociation) found on most commercial mass spectrometers. Threshold dissociation mechanisms for protein dissociation are known to produce b- and y-type fragment ions (fragments containing the amino and carboxy terminus, respectively); however, backbone cleavage often occurs at preferred amino acids in the primary sequence, resulting in diminishing sequence information as protein size increases. Additionally, the energy imparted by threshold methods is known to selectively eject labile posttranslational modifications such as glycosylation, thereby producing very little sequence information and/or localization information of these modifications to particular amino acid residues (15). Today, several commercial instruments are equipped with electron capture (ECD) (16, 17) or electron-transfer dissociation (ETD) (18) capabilities, which induce low-energy, nonselective backbone fragmentation to produce c- and z-type fragment ions (fragments containing the amino and carboxy terminus, respectively). This allows for extensive gas-phase sequencing while simultaneously preserving and localizing modifications. Lastly, the ideal protein identification algorithm would not only identify the alternative fragmentation products but accommodate consistent mass shifts due to both known and unknown posttranslational modifications and/or truncations and deletions. Software tools, such as ProSightPC (Thermo Scientific), provide thorough analysis of top-down, high-resolution tandem mass spectrometry data for protein identification and characterization (19, 20).

Top-down dMS is an ideal approach towards understanding the subtle protein changes that define the effect of HDL particles. Higher levels of HDL cholesterol (HDL-c) are understood to be generally cardioprotective, although use of HDL-c levels as a biomarker is not always accurate in predicting atherosclerotic risk (21). Several groups are applying proteomics techniques to understand the protein composition that define HDL function using peptide-counting methods for quantitation (22–24). In addition to the major protein contribution of apolipoprotein A-I and apolipoprotein A-II molecules (22, 25), many of the proteins identified from HDL particles are relatively low in molecular weight (most below 30 kDa). This relatively simple and low molecular weight protein mixture is well suited for top-down mass spectrometry. The experiment described here includes the dMS analysis of intact lipoproteins from samples comparing a subclass of human HDL, namely, HDL_3 (d = 1.13–1.21), from subjects having relatively high and low HDL-c levels. Alternatively, samples of total HDL (d = 1.07–1.21) apolipoproteins obtained from donors with high or low LDL (low-density lipoprotein) cholesterol could be analyzed to better understand the relationship between LDL-c and HDL protein composition. However, the comparison chosen here of HDL_3 apolipoproteins obtained from subjects with high and low total HDL-c presents a basic 2-factor comparison for simplicity and is used to shed light on the relationship between HDL_3 protein composition and HDL-c levels.

An ultracentrifugation method was used for isolation of the HDL_3 subfraction from fresh EDTA-plasma obtained from three subjects containing relatively low HDL-c levels (average: 44.3 mg/dL) and three subjects containing relatively high HDL-c levels (average: 73.7 mg/dL). The three biological replicates from each high and low HDL-c levels were further aliquoted to three technical replicates, yielding nine samples for each factor. All samples were processed identically by sample desalting (which also removes the majority of lipids) and reduction. Samples were analyzed by LC-MS profiling using a hybrid high-resolution Fourier-transform ion cyclotron resonance mass spectrometer (i.e., LTQ-FT Ultra). Elucidator was used for chromatographic time alignment, feature detection and grouping, and statistical analysis. An LTQ Orbitrap XL with ETD was subsequently employed for selected MS/MS targeting of features of interest, and resulting tandem mass spectra were submitted to ProSightPC for data analysis and protein characterization.

It is worth noting that the quantitative, label-free platform detailed here is highly biased towards the analysis by high-resolution instruments (particularly, Thermo Scientific LTQ-FT and Orbitrap instruments) and Elucidator data processing. Although high-performance instruments are continuously being developed, it is vital that any instruments used provide very high mass accuracy (i.e., <5 ppm) and resolution (>60,000). This is critical for both isotopic resolution of highly charged, intact protein species and for feature extraction by dMS analysis. The attractiveness of Elucidator lies in its powerful ability to normalize and align LC-MS "images," accurately extract features based on spectral metrics as opposed to MS/MS identification, perform statistical tests, and manually inspect/visualize results. If Elucidator is not feasible for use, scientists can employ other available quantitative, label-free software packages, provided the feature extraction algorithms used are accurate and reliable. This software should be pressure tested against highly charged analytes to ensure proper peak assignment, grouping, and chromatographic time alignment. Lastly, one can employ more commonly accessible software tools (e.g., Microsoft Excel) for the statistical analysis and data manipulation (detailed in this chapter to assist the reader).

2 Materials

All reagents used for the preparation and analysis of human HDL_3 should be of HPLC-grade quality or better to ensure that no low molecular weight contaminants are introduced into the analysis (*see* **Note 1**).

It is critical that all samples be prepared in parallel with identical reagents/procedures to ensure consistent sample processing required for dMS analysis (*see* **Note 2**).

Store all reagents at room temperature unless otherwise noted.

2.1 HDL Preparation

1. HDL_3 samples (at least 6 total), obtained from donors with high (3) and low (3) HDL cholesterol (HDL-c): prepared according to Chapman et al. (26), (*see* **Note 3**).
2. BCA assay kit, Thermo Scientific, Product #23255.
3. UV/Visible spectrophotometer.
4. C_4 ZipTip desalting tips (Product #: ZTC04S008) (*see* **Note 4**).
5. SpeedVac concentrator: Thermo Scientific Savant SDP1010, or equivalent.
6. Mass spectrometry compatible centrifuge tubes and autosampler vials.

2.2 LC-MS Reagents

1. HPLC solvent A (0.1 M acetic acid): add 25 mL HPLC-grade acetic acid directly to 4 L of HPLC-grade water. Invert several times to ensure adequate mixing.
2. HPLC solvent B (0.1 M acetic acid in acetonitrile): add 25 mL HPLC-grade acetic acid directly to 4 L of HPLC-grade acetonitrile. Invert several times to ensure adequate mixing.
3. Instrument quality assurance/quality control (QA/QC) sample: prepare 100 fmol/μL of each bradykinin (Sigma Aldrich Product #: B3259), angiotensin (Sigma Aldrich Product #: A9650), and neurotensin (Sigma Aldrich Product #: N6383) in HPLC solvent A.
4. Custom-packed capillary HPLC columns: POROS 1 stationary phase packed into 75 μm I.D. × 6 cm fused-silica capillary (*see* **Note 5**).
5. Fused-silica capillary tubing, 75 μm ID (360 μm O.D.) (Polymicro Technologies Product #: TSP75375), approximately 1 m in length.
6. Cross for 360 mm OD fused-silica tubing (Chromtech Product #: P-888), tubing sleeves (Chromtech Product #: F-185X), and plug (Chromtech Product #: P-551).
7. Halo® C8 trap column (New Objective, Item #: 25H036) and holder (New Objective, Item #: TRC-25).
8. HPLC system: Agilent HP1200 capillary pump and autosampler.
9. Thermo Scientific Nanospray Flex Ion Source (Product #: ES071).
10. Thermo Electron LTQ-FT Ultra mass spectrometer.
11. Thermo Electron LTQ Orbitrap XL mass spectrometer with ETD capabilities.

3 Methods

3.1 Preparation of Human High-Density Lipoprotein Samples

1. Prepare human HDL_3 particles from fresh EDTA-plasma by density gradient ultracentrifugation (d=1.13–1.21 g/mL) using established methods (26) (*see* **Note 6**).
2. Measure total protein concentrations for prepared HDL_3 particles by bicinchoninic assay (BCA) to ensure consistent sample preparation (*see* **Note 7**).
3. Aliquot approximately 3 μg (~3 μL of 1 mg/mL sample) of total protein from each HDL_3 sample to fresh centrifuge tube and add 10 μL of 0.1 M acetic acid. Desalt samples using ZipTips according to the manufacturer's protocol (*see* **Note 8**).
4. Repeat the aliquot/ZipTip desalting process two additional times to generate three technical replicates for each HDL_3 sample (*see* **Note 9**). Label appropriately.
5. Take desalted samples (18 in total) to dryness in the SpeedVac; store dry at −20 °C until LC-MS analysis.

3.2 Preparation for LC-MS Analysis

1. Create an LC-MS sequence list in which HDL_3 samples from different conditions (i.e., HDL_3 isolated from donors with high and low HDL-c levels) are interwoven throughout the run list (*see* **Note 10**), with instrument QA/QC samples (*see* **Note 11**) included at least every six analysis samples. See Fig. 2 as an example.
2. Prepare and operate the HPLC pump and autosampler plumbing according to Fig. 3 (*see* **Note 12**). Use the Thermo Scientific Nanospray Flex Ion Source for electrospray ionization of LC column eluent into the LTQ-FT Ultra mass spectrometer (*see* **Note 13**).
3. Prepare the HPLC with appropriate mobile phase (i.e., solvent A and solvent B) and purge solvent lines to ensure no air bubbles are present.
4. Ensure the proper performance of HPLC and mass spectrometry instrument (i.e., LC-MS) by analyzing with instrument QA/QC controls prior to beginning the experiment. Ensure proper instrument mass accuracy (<5 ppm), resolution (>60,000 at 433 m/z), and signal intensity (typically base peak LC-MS ion intensity of >1e6 for LTQ-FT Ultra instrument).

3.3 LC-MS Analysis of Prepared HDL_3 Samples

1. Prepare each HDL_3 sample for analysis by resuspending in 10 μL of 0.1 M acetic acid containing 10 mM dithiothreitol (DTT) (*see* **Note 14**). Vortex samples and centrifuge at 14,000×g for 3 min (*see* **Note 15**). Carefully aliquot samples to fresh autosampler vials and place in appropriate location of the autosampler.

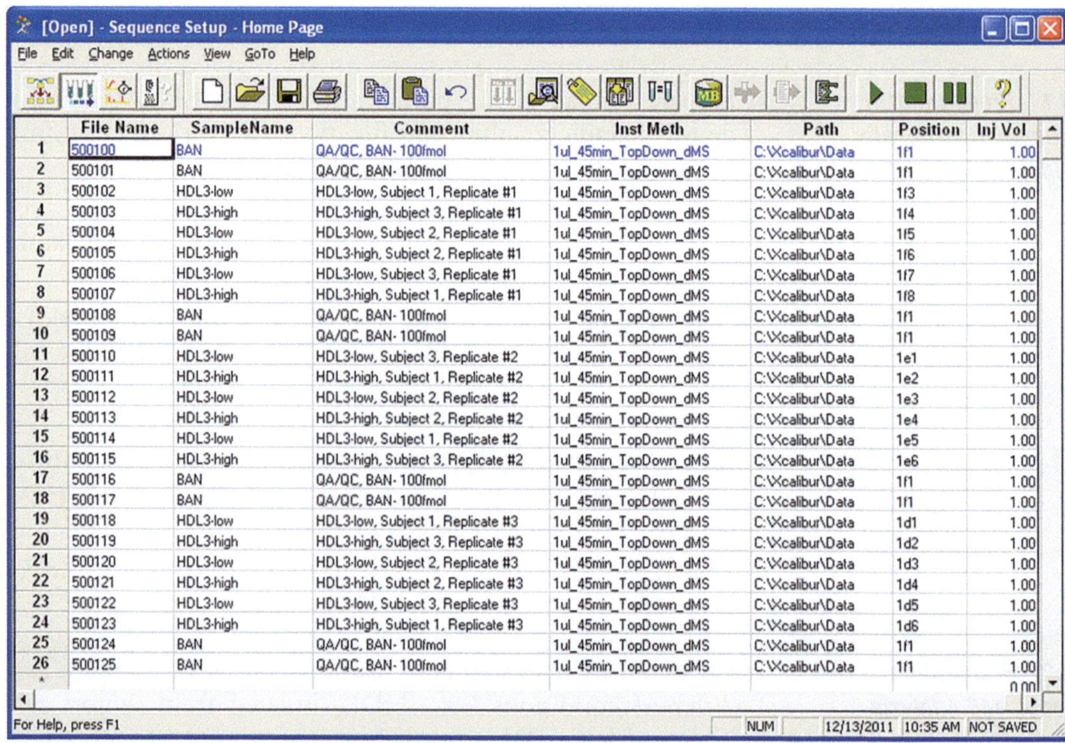

[Open] - Sequence Setup - Home Page

File Edit Change Actions View GoTo Help

	File Name	SampleName	Comment	Inst Meth	Path	Position	Inj Vol
1	500100	BAN	QA/QC, BAN- 100fmol	1ul_45min_TopDown_dMS	C:\Xcalibur\Data	1f1	1.00
2	500101	BAN	QA/QC, BAN- 100fmol	1ul_45min_TopDown_dMS	C:\Xcalibur\Data	1f1	1.00
3	500102	HDL3-low	HDL3-low, Subject 1, Replicate #1	1ul_45min_TopDown_dMS	C:\Xcalibur\Data	1f3	1.00
4	500103	HDL3-high	HDL3-high, Subject 3, Replicate #1	1ul_45min_TopDown_dMS	C:\Xcalibur\Data	1f4	1.00
5	500104	HDL3-low	HDL3-low, Subject 2, Replicate #1	1ul_45min_TopDown_dMS	C:\Xcalibur\Data	1f5	1.00
6	500105	HDL3-high	HDL3-high, Subject 2, Replicate #1	1ul_45min_TopDown_dMS	C:\Xcalibur\Data	1f6	1.00
7	500106	HDL3-low	HDL3-low, Subject 3, Replicate #1	1ul_45min_TopDown_dMS	C:\Xcalibur\Data	1f7	1.00
8	500107	HDL3-high	HDL3-high, Subject 1, Replicate #1	1ul_45min_TopDown_dMS	C:\Xcalibur\Data	1f8	1.00
9	500108	BAN	QA/QC, BAN- 100fmol	1ul_45min_TopDown_dMS	C:\Xcalibur\Data	1f1	1.00
10	500109	BAN	QA/QC, BAN- 100fmol	1ul_45min_TopDown_dMS	C:\Xcalibur\Data	1f1	1.00
11	500110	HDL3-low	HDL3-low, Subject 3, Replicate #2	1ul_45min_TopDown_dMS	C:\Xcalibur\Data	1e1	1.00
12	500111	HDL3-high	HDL3-high, Subject 1, Replicate #2	1ul_45min_TopDown_dMS	C:\Xcalibur\Data	1e2	1.00
13	500112	HDL3-low	HDL3-low, Subject 2, Replicate #2	1ul_45min_TopDown_dMS	C:\Xcalibur\Data	1e3	1.00
14	500113	HDL3-high	HDL3-high, Subject 2, Replicate #2	1ul_45min_TopDown_dMS	C:\Xcalibur\Data	1e4	1.00
15	500114	HDL3-low	HDL3-low, Subject 1, Replicate #2	1ul_45min_TopDown_dMS	C:\Xcalibur\Data	1e5	1.00
16	500115	HDL3-high	HDL3-high, Subject 3, Replicate #2	1ul_45min_TopDown_dMS	C:\Xcalibur\Data	1e6	1.00
17	500116	BAN	QA/QC, BAN- 100fmol	1ul_45min_TopDown_dMS	C:\Xcalibur\Data	1f1	1.00
18	500117	BAN	QA/QC, BAN- 100fmol	1ul_45min_TopDown_dMS	C:\Xcalibur\Data	1f1	1.00
19	500118	HDL3-low	HDL3-low, Subject 1, Replicate #3	1ul_45min_TopDown_dMS	C:\Xcalibur\Data	1d1	1.00
20	500119	HDL3-high	HDL3-high, Subject 3, Replicate #3	1ul_45min_TopDown_dMS	C:\Xcalibur\Data	1d2	1.00
21	500120	HDL3-low	HDL3-low, Subject 2, Replicate #3	1ul_45min_TopDown_dMS	C:\Xcalibur\Data	1d3	1.00
22	500121	HDL3-high	HDL3-high, Subject 2, Replicate #3	1ul_45min_TopDown_dMS	C:\Xcalibur\Data	1d4	1.00
23	500122	HDL3-low	HDL3-low, Subject 3, Replicate #3	1ul_45min_TopDown_dMS	C:\Xcalibur\Data	1d5	1.00
24	500123	HDL3-high	HDL3-high, Subject 1, Replicate #3	1ul_45min_TopDown_dMS	C:\Xcalibur\Data	1d6	1.00
25	500124	BAN	QA/QC, BAN- 100fmol	1ul_45min_TopDown_dMS	C:\Xcalibur\Data	1f1	1.00
26	500125	BAN	QA/QC, BAN- 100fmol	1ul_45min_TopDown_dMS	C:\Xcalibur\Data	1f1	1.00
*							0.00

For Help, press F1 NUM 12/13/2011 10:35 AM NOT SAVED

Fig. 2 Example of LC-MS sequence run list for a dMS experiment to compare HDL proteins from subjects having high and low HDL-c levels. The samples are interwoven by condition to reduce or eliminate any systematic bias that may result from LC-MS analysis

2. Create an LC-MS method for the analysis, and use the LC gradient method conditions according to Table 1 (*see* **Note 16**). Ensure that the tune file used is optimized for intact protein analysis (*see* **Note 17**).
3. Loop-inject samples (1 μL, or ~10 pmol total protein on column) onto Halo® C8 trap column (1 cm × 75 μm) using the Agilent HP1200 capillary autosampler. Wash the trapped analytes at 3 μL/min with 90 % 0.1 M acetic acid (solvent A) and 10 % 0.1 M acetic acid in acetonitrile (solvent B) for 5 min to remove any residual salt or hydrophilic components. At 5.1 min, reduce flow to 1 μL/min and direct flow from waste to the instrument (i.e., source). A gradient is then performed to separate intact protein components (*see* Table 1 and **Note 18**).
4. Collect only full-scan high-resolution spectra (i.e., no tandem mass spectrometry experiments), approximately once every 1.5 s throughout the LC-MS analysis (*see* **Note 19**).
5. Analyze all HDL_3 samples by reverse-phase liquid chromatography coupled to a LTQ-FT Ultra mass spectrometer (*see* **Note 20**). Collect all sample data without interruption of the designed LC-MS sequence (*see* **Note 21**).

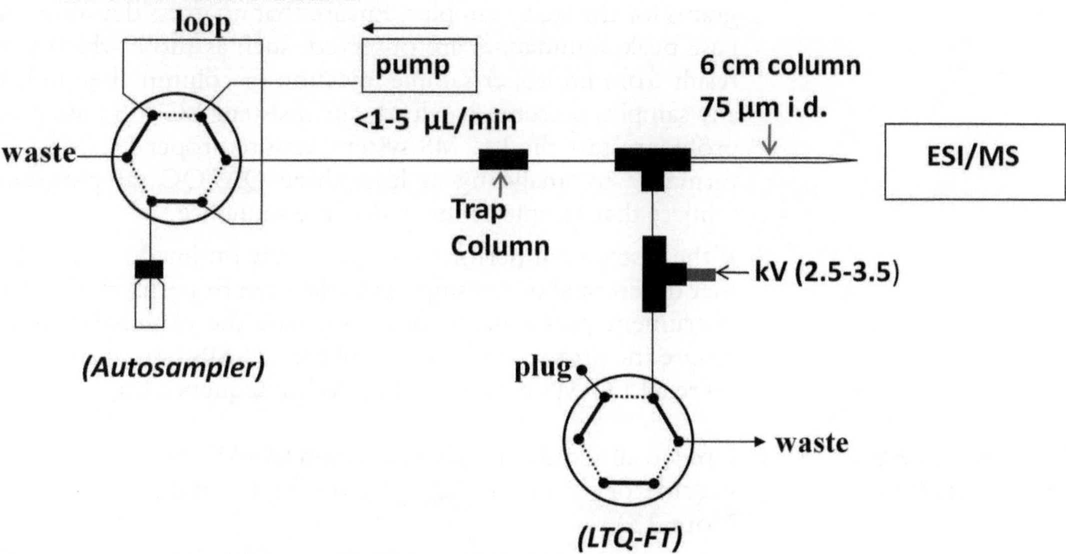

Fig. 3 LC-MS setup for dMS experiment. The autosampler loop-injects the sample and is further trapped on the C8 trap column. The flow is directed through the trap column to waste to allow sample desalting. Following the 5 min wash, the flow is directed through the analytical capillary LC column by switching the divert valve to plug the waste flow. This arrangement allows both an offline sample desalting (i.e., flow to waste) and an indirect voltage connection for electrospray ionization

Table 1
Liquid chromatography method conditions

Time	Solvent A (%)	Solvent B (%)	Flow (µL/min)	Divert valve
0	90	10	3	Waste
5	90	10	3	Waste
5.1	90	10	1	Source
30	55	45	1	Source
35	10	90	1	Source
39	10	90	1	Source
39.1	90	10	3	Waste
45	90	10	3	Waste

6. Evaluate the analytical consistency of the acquired data using the included QA/QC samples by determining the retention time, peak area, and mass accuracy of each bradykinin, angiotensin, and neurotensin peptides (*see* **Note 22**). For acceptance, we required that percent coefficient of variation (%CV) for the peak area be <10 % and the mass accuracy and retention time not drift by more than 15 ppm and 1 min, respectively, over the course of the experiment.

7. Qualitatively inspect the acquired LC-MS base peak chromatograms for the study samples. Ensure that no gross deviations in base peak abundances are observed, such as those which may result from improper sample injection or column clogging. If any samples presented a clearly inconsistent LC-MS base peak profile, restart the LC-MS system, ensure proper LC-MS performance by analyzing at least three QA/QC samples, and reinject that sample at the end of the sequence.
8. If the instrument performance passes the outlined criteria, further differential processing of the data can be performed. If the instrument performance does not pass the outlined criteria, ensure the proper performance of the LC-MS system and reacquire all LC-MS data according to the sequence list.

3.4 dMS Analysis of HDL_3 Samples

1. Upload all acquired high-resolution LC-MS data into Rosetta Elucidator (version 3.2, Microsoft) for data analysis (*see* **Note 23**).
2. Using the PeakTeller™ algorithm within Elucidator, align LC-MS data sets in time, normalize for intensity, and extract "features" (*see* **Note 24**). To accomplish this, create an Experimental Definition (ED) containing the representative-acquired LC-MS dataset and group the data files according to condition (e.g., HDL_3 from donors with high and low HDL-c) (*see* **Note 25**). Run the Experimental Definition.
3. Following feature extraction, qualitatively inspect feature fold-change data (i.e., ratio data). The majority of features should have a ratio between conditions of approximately 1 or not be statistically significant via Elucidator p-value <0.01 (*see* Fig. 4 and **Note 26**).
4. Select features containing area under curve (AUC) values greater than zero for all samples within one given condition for further statistical testing (*see* **Notes 27** and **28**).
5. Export the selected feature attributes (m/z, retention time, AUC, mass, unique feature ID, etc.) to a suitable spreadsheet program (e.g., Microsoft Excel) for processing (*see* **Note 29**).
6. Average replicate feature AUC's from each condition (HDL_3 from donors with high and low HDL-c).
7. Perform a two-sample t-test of the averaged biological and technical replicates using the logarithmic of average AUC (i.e., log AUC) (*see* **Note 30**).
8. Filter features by a user-defined level of significance (typically p-value <0.02). We require that at least two features per isotope group (i.e., features from the same isotopic distribution) demonstrate statistical significance.
9. Significant features can be further filtered based on molecular weight, charge state range, and/or retention time (*see* **Note 31**).

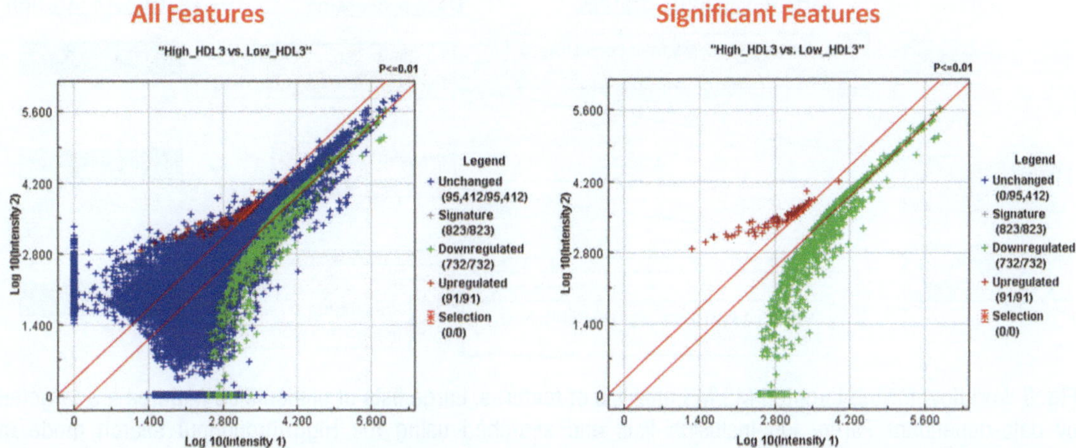

Fig. 4 Ratio data of feature abundances comparing subjects with high and low HDL-c levels. Ratio data plot compares the feature log intensity of condition 1 (HDL from subjects with high HDL-c) to log intensity of condition 2 (HDL from subjects with low HDL-c). 45° line represents intensity ratio between conditions equal to one (i.e., equal intensity between both conditions). Plots show all features (*left*) and only those determined statistically significant via Elucidator significance testing, $P < 0.01$ (*right*)

10. Select statistically significant features of interest for further tandem mass spectrometry analysis (*see* **Note 32**).

3.5 Targeted Analysis and Online LC-MS/MS of Features of Interest

1. For smaller lists of features to be targeted (<20 total), create targeted MS/MS methods in Xcalibur in which dedicated scan events are created to collect high-resolution ETD and/or CID spectra for specific accurate *m/z* values over the entire course of an established LC segment (determined based on retention time ± appropriate window) (*see* **Note 33**). For larger lists of features for targeting (>20 total), create targeted MS/MS methods in Xcalibur which utilize parent ion inclusion lists to selectively target detected precursor ions from full-scan spectra (*see* **Note 34**).
2. Keep chromatographic conditions for the targeted MS/MS method consistent with that from the dMS profiling method (*see* **Note 35**).
3. Use a tune file in which CID and ETD had been optimized for the average precursor charge state on the targeted feature list (*see* **Note 36**).
4. Choose one sample from each condition (i.e., high and low HDL-c levels) and analyze both selected samples by targeted MS/MS methods (above).
5. Collect ETD and CID spectra in the Orbitrap mass analyzer with an instrument-resolving power of at least 15,000. Ensure that scan range for the ETD spectra is *m/z* 100 to 2,000.
6. For the targeted MS/MS data files from smaller list of features (<20), extract each targeted precursor ion (full-scan spectra),

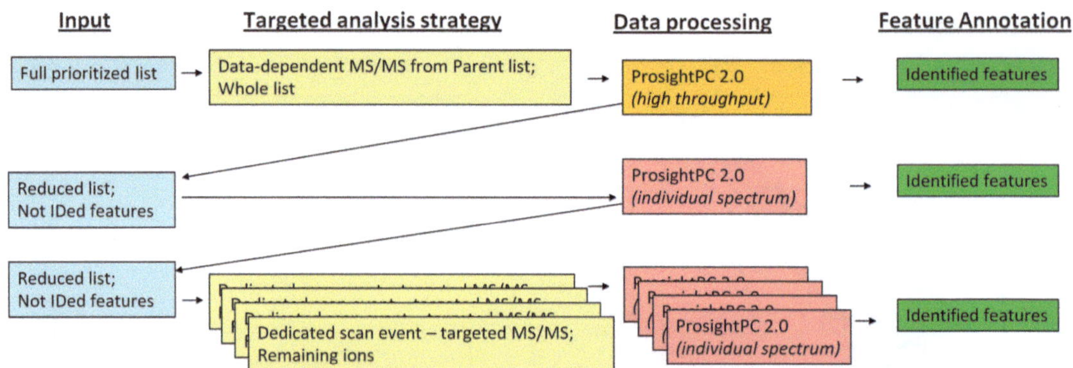

Fig. 5 Workflow for the targeted MS/MS analysis of features. Large lists of unidentified features are targeted by data-dependant Parent ion inclusion lists and searched using the High-throughput search mode of ProSightPC. Those features readily identified are annotated and removed from further analysis. Those features not identified are further analyzed by the individual spectrum analysis functionality of ProSightPC, and identified features are again annotated and removed. The remaining features are then selectively targeted by MS/MS experiments, and analyzed by ProSightPC to annotate those features not previously identified by Parent Ion inclusion list experiments

identify the corresponding MS/MS spectra (ETD or CID), and manually sum the fragmentation spectra across the peak elution. Save the product spectrum as a separate file (*see* **Note 37**). For the targeted MS/MS data files from the larger list of features (>20), analyze the raw LC-MS/MS data directly in ProSightPC v2.0 using high-throughput mode (see below).

7. Analyze data using ProSightPC v2.0.

3.6 Protein Identification of Targeted Features by ProSightPC v2.0

1. Both individual spectra (i.e., single spectrum) and entire LC-MS/MS data can be processed by ProSightPC v2.0. For the targeted MS/MS analysis of smaller lists of features (<20 features), analyze the data using single spectrum search mode. For the targeted MS/MS analysis of larger lists of features (>20 features), use the high-throughput function of ProSightPC v2.0 (Fig. 5).
2. To search individual MS/MS spectra, create an Experiment (i.e., Build Experiment from Profile Raw Data) for each saved MS/MS spectra, de-isotope the spectrum, and perform an Accurate Mass Search (*see* **Note 38**). If the search results return proteins identified with *P*-score and *E*-score values of <1E−5, manually verify the results for accuracy (*see* **Note 39**). If the search does not return proteins with P-score and E-score values of <1E−5, repeat the search in Biomarker Mode (*see* **Note 40**). If the search does not return proteins with P-score and E-score values of <1E−5, the spectrum will likely need to be interpreted manually (*see* **Note 41**).

3. To search entire LC-MS/MS data files, use the High-throughput mode of ProSightPC v2.0 (*see* **Note 42**). Create an appropriately named data repository for the search results and perform the search.
4. Search results can be visualized and sorted by opening the respective data search Repository Report. All appropriate search attributes are displayed, including protein identity, *P*-score, *E*-value, observed and theoretical matches and number of matching fragments. Filter those proteins having *P*-score and *E*-score values of <1E−5.
5. Ensure the matched protein identification search results are the desired targeted features by comparing the identified mass with the accurate mass of each feature. Further inspect that the feature retention time within the raw MS/MS data file is at the same approximate retention time as that observed for the original dMS experiment.
6. Annotate identified features appropriately according to the software being used (*see* **Note 43**).

4 Notes

1. Reducing the amount of low molecular weight contaminants reduces the number of features derived from such contaminants. This greatly aids in the analysis of bioinformatic data and reduces false-positive features.
2. The foundation of a successful dMS experiment is the reproducible sample preparation and LC-MS analysis of samples. Artificially introduced deviations will result in false-positive "differences," obscuring authentic protein abundance differences. Utmost care should be taken to prepare all samples identically (without bias) and analyzed in an interwoven design to eliminate systematic bias due to run order.
3. Human HDL_3 subfraction purified by Sequential Density Ultracentrifugation can also be purchased from several vendors, including GenWay Biotech, Inc. (Product #: GWB-2FA11B). Ensure the proper HDL-c attributes are known and selected.
4. We use ZipTip sample preparation for desalting small amounts of material (typically <5 μg); however, HPLC columns or trap columns (e.g., Michrom Bioresources, Protein MacroTrap, Product #: TR1/25108/53) can be used in replacement.
5. If in-house custom packing of capillary LC columns is not possible, several vendors (e.g., New Objective, Mcllent) offer custom packing services for capillary LC columns.

6. HDL_3 should be stored at 4 °C when not being used. Although no published reports describe a perturbation in protein stability when stored at −20°C or −80 °C, frozen HDL_3 particles may be sensitive to changes that may affect the overall protein compositions and, hence, differential analysis.
7. Care should be taken to determine the total protein concentration of each prepared HDL_3 sample, as any gross deviation in protein amounts between samples can challenge the alignment, normalization, and feature extraction of the dMS algorithm. Protein concentrations can be determined using bicinchoninic assay kits (http://piercenet.com/instructions/2161296.pdf).
8. ZipTip desalting is done according to manufacturer's protocol (http://www.millipore.com/userguides/tech1/pr02358). This preparation removes salts and depletes the sample of very hydrophobic analytes (e.g., lipids).
9. The creation of technical replicates is used to increase the statistical power of the experiment. Ideally, additional biological replicates (i.e., HDL_3 isolated from different donors with the same high or low HDL-c condition) would improve the (biological) confidence in the statistical results.
10. We create an LC-MS sequence run list by interweaving samples from different conditions throughout the running of the sequence. This should reduce any systematic bias from LC-MS analysis.
11. Any suitable instrument QA/QC sample able to adequately measure performance can be used. We used a peptide mixture of bradykinin, angiotensin, and neurotensin (each at 100 fmol/μL), with known masses and retention times. Tracking of instrument mass deviations and retention time drifts provide a measure of mass accuracy and chromatographic alignment tolerance, respectively.
12. The LC plumbing uses a second switching valve at the LTQ-FT Ultra mass spectrometer to direct flow either through the capillary LC column (i.e., switching valve to plug) or to waste (i.e., switching valve to waste). This allows for sample desalting via trap column (switch valve directed to waste) and gradient elution of protein components (switch valve directed to plug). We found this column arrangement beneficial for both consistent chromatography and electrospray ionization (i.e., indirect application of electrospray voltage).
13. Because of the relatively low chromatographic flow rates (1 μL/min), we use the nanospray source (i.e., Thermo Scientific Nanospray Flex Ion Source) as the electrospray ionization assembly instead of the standard Ion Max Source (used for higher flow rates, typically 200–550 μL/min). We found that having a nano- to microscale LC with direct mass

spectrometer interface, without flow splitting, provides optimum sensitivity and chromatographic reproducibility.

14. Tris(2-carboxyethyl)phosphine (TCEP)-related chemical cleavage has recently been reported (27). As a result, dithiothreitol (DTT) should be used in replacement to reduce any disulfide bonds which may be present within the protein mixtures.

15. We ensure that the samples are adequately centrifuged to remove (i.e., pellet) any residual insoluble particles which may affect the chromatographic backpressure and, hence, retention time. After centrifugation, carefully transfer the liquid to a fresh autosampler vial.

16. For the mass spectrometry method, we use the following conditions: one scan event, full-scan FT detection, scan range 300–2,000 m/z, profile mode, and resolving power 100,000. Profile mode format of raw LC-MS data is required for many quantitative, label-free software algorithms including Elucidator PeakTeller™ and dMS.

17. We modified our instrument conditions for intact protein analytes to include 3 kV electrospray voltage, heated metal capillary temperature 270 °C, a tube lens of 80 V, ion accumulation target value (determined by the instrument automatic gain control, AGC) of 1E7, and max injection time of 1 s. We found other tune parameters are less sensitive to intact protein detection and can be determined by the instrument automatic tuning using an intact protein calibrant (e.g., bovine ubiquitin).

18. Reproducible, stable chromatography is critical for successful differential analysis. In our experience, it was found to be more beneficial to have consistent chromatographic separations, at the expense of high-resolution chromatography. In fact, top-down dMS analysis provided the most reliable quantitative data from chromatographic peak widths of >30 s. We employed relatively low performance POROS™ stationary material (within the flow rate used) and relatively simple solvent gradient as these conditions were optimum for the successful feature extraction and relative quantitation.

19. Because high-resolution tandem mass spectrometry experiments reduces the instrument duty cycle (i.e., number of data points collected across the chromatographic peak), we performed only full-scan LC-MS profiling experiments. Tandem mass spectrometry experiments were performed following the dMS experiment.

20. Conceptually, any high-resolution mass spectrometry instrument can be used for dMS analysis. Because isotopic resolution of protein analytes is required for dMS analysis, use of an instrument capable of producing data with >60,000 resolution

is highly desired. One should note, however, that certain instruments such as the LTQ-Orbitrap will have inherent upper mass limits due to their means of ion transmission, despite their ability to isotopically resolve large, highly charged ions at the point of detection. This should be considered when choosing a high-resolution mass spectrometer for quantitative profiling at a given mass range.

21. We found that collecting all LC-MS data start to finish without interruption maintains the consistency of the chromatography. Additionally, all LC-MS data should be collected using a single trap and HPLC capillary column setup (i.e., we do not change any HPLC hardware during LC-MS acquisition). Abrupt stops or pauses of the sequence can affect the retention time consistency and the final dMS analysis. If the LC-MS sequence inadvertently stops during the data acquisition, restart the LC-MS system, allow 3–4 QA/QC injections, and then proceed with the sequence file.

22. The peak area and retention time for each of the bradykinin, angiotensin, and neurotensin peptides should be obtained by extracted ion chromatograms specific for each respective peptide via Xcalibur. To do this, the ions 530.7880 m/z (bradykinin), 432.8998 m/z (angiotensin), and 558.3104 m/z (neurotensin) are used as inputs for range(s) to be extracted, using a user-defined mass tolerance of 10 ppm. Peak area (area under curve, AUC) is determined by toggling peak detection in Xcalibur.

23. Although multiple software tools that can handle the computational demands of quantitative, label-free proteomics experiments exist (e.g., Rosetta's Elucidator, Thermo Scientific's SIEVE, Waters' BioPharmaLynx), we have found Elucidator's ability to accurately identify features (as an accurate m/z, retention time, and intensity) is essential for our dMS experiments. In addition to the accurate feature identification via the PeakTeller™ algorithm, Elucidator provides a host of software solutions we found highly useful, including data viewers, search engine integration, verification viewers, customizable statistical methods, and accommodation of raw data from most instrument vendors. Other software packages may provide similar solutions, although none (other than Elucidator) have been tested in our lab. For a detailed review of existing quantitative, label-free software programs, see Mueller et al. (28).

24. We found Elucidator's PeakTeller™ algorithm to be highly reliable with respect to raw data processing, including chromatographic time alignment, intensity normalization, peak detection, charge-state determination, feature quantitation, and feature quality scoring. PeakTeller™ also has the ability to accurately cluster features into their corresponding isotope groups (i.e., features from the same isotopic distribution of

one protein species) and charge groups (i.e., features form the same isotopic distribution from the multiple charge states of one protein species), based on accurate mass, charge, and isotope pattern recognition. These characteristics make the software extremely useful and enable the user the flexibility to apply customizable filters for feature extraction, prioritization, and MS/MS identification.

25. The parameters we used for Experimental Definition creation were Design Type = Differential-Label Free; Add Factor = condition (add two levels, high HDL-c and low HDL-c); Select a Common Filter = FTMS; nine raw data files assigned to each treatment group (i.e., high HDL-c and low HDL-c); process all Time and *m/z* ranges; and Build using = PeakTeller (FT). Aligned Data Advanced parameters to include Peak Time and Peak *m/z* score = 0, Peak Time width = 0.1 min, Calculate feature intensity = Height, Intensity threshold cutoff = 0, Alignment algorithm = Spectral Alignment, instrument mass accuracy = 10 ppm, Alignment search distance = 4 min.; default/automatic Filtering, Feature Identification, and Feature Removal; Max charge state for grouping = 30. Ratio Builder Advanced parameters to allow for Perform Scaling using Mean; Statistical Cut Trim percentage = 10; and Build Feature and Isotope Level Ratios using the Generic Ratio Error Model with a fold-change limit to 1,000.

26. Because of the relative protein composition similarity between samples, it is expected that the majority of features should not be significantly different between conditions (i.e., features' ratio data should be approximately ~1.0). If the majority of features are skewed towards one condition, this is an indication of a potential problem in the dMS experiment. We have found that often this is a result of low-quality LC-MS data file, as is the case for an inaccurate injection amount (e.g., the sample volume not at the bottom of the autosampler vial during sample aspiration). If Elucidator is not used for feature extraction, qualitatively compare the relative abundances of features between conditions to ensure the majority are clustered around a value of 1.0.

27. We find the extensive feature details (e.g., feature ID, feature *m/z*, retention time, AUC, fold-change), powerful Boolean searching, and data exporting (.csv file) capabilities of Rosetta Elucidator extremely advantageous for downstream data processing.

28. The procedure for selecting those features which display a statistically significant abundance difference between conditions requires careful filtering of feature data. In order to enrich features for those ions which would be considered authentic protein species (i.e., not false-positive features that result from

noise or interference), we empirically found that only those ions containing AUC values greater than zero for all samples within one given condition should be considered for further significant testing.

29. Following the filtering of features according to AUC, only a relatively small subset of total features will remain. Data manipulation and significance testing can be accomplished with a simple spreadsheet program that allows simple cell calculations and sorting mechanisms.

30. The calculation of a two-tailed t-test is performed using the equation

$$t = \frac{\overline{X}_1 - \overline{X}_2}{s_{\overline{X}_1 - \overline{X}_2}}$$

where

$$s_{\overline{X}_1 - \overline{X}_2} = \sqrt{\frac{s_1^2}{n_1} + \frac{s_2^2}{n_2}}$$

and $\overline{X}_1$ and $\overline{X}_2$ are the mean log AUC of conditions 1 (i.e., high HDL-c subjects) and 2 (i.e., low HDL-c subjects), respectively; s_1 and s_2 are the standard deviation of the log AUC of conditions 1 (i.e., high HDL-c subjects) and 2 (i.e., low HDL-c subjects), respectively; and n_1 and n_2 equal to the number of samples analyzed for each conditions 1 and 2, respectively (n_1 and $n_2 = 3$, for this experiment). Once a t-value is calculated, a probability can be determined via Student's t-distribution. If more complicated experiments are to be performed (i.e., additional conditions or variables), a more powerful software program (e.g., Elucidator) should be used.

31. The number of features generated by a dMS experiment is determined by many factors, including the design of the experiment (i.e., comparison of very similar or different samples, the number of factors, etc.), the complexity of the sample mixture, the quality of sample processing, the quality of LC-MS acquisition, and the feature extraction parameters. We found that filtering out features by molecular weight (<3 kDa), charge state (<4+ charge), and elution time (<5 or >40 min) reduced the amount of features of interest and focus the results to intact proteins. Additionally, features from the same protein analytes can be further "collapsed" into isotope groups and charge groups to eliminate redundant features from the same protein. Lastly, we prioritized features based on fold change between conditions and p-value to determine those features of most interest to the experiment.

32. If Elucidator is not used, careful inspection of the statistically significant features can greatly reduce the number of features

required for targeted MS/MS analysis and identification. Therefore, it is often advantageous to "front filter" those features not of interest (or redundant via isotope or charge grouping) prior to expending the laboratory resources for protein identification.

33. We use an LC time segment of at least ±1.5 min of the feature retention time determined from the dMS experiment. This ensures the maximum likelihood of collecting at least one MS/MS spectrum for each targeted feature.

34. We found the use of data dependent MS/MS methods using Parent ion inclusion lists an efficient way to quickly and selectively identify a number of ions from the prioritized list in a single pass. To do this, one full-scan FTMS scan event was followed with 3–5 data-dependant high-resolution ETD scan events. The Parent ion inclusion list was populated with the targeted features *m/z*, and the FTMS/MS events directed to target the ions from the Parent ion list. If an ion from the list is detected above the set threshold (typically, 1E4), an MS/MS event is triggered even if this ion is not among the most abundant peaks in the spectrum. In this way, ions of interest can be selectively fragmented against a complex background in an automated fashion. If no ions from the list are detected above threshold, MS/MS is not triggered and the instrument returns to the next FTMS full scan, maintaining optimum duty cycle. We used the following method parameters: parent ion mass tolerance to 10 ppm, dynamic exclusion repeat count of 2–5 (depending on list size), exclusion duration to 60 s, and exclusion list size to 500. Each triggered MS/MS was collected with three microscans, a max fill time of 1,200 ms, and a precursor target value of 1E5. We used an Orbitrap MS/MS resolving power of 15,000 to insure isotopic resolution of most fragments.

35. Because the chromatographic conditions used for targeted MS/MS are identical to the original dMS experiment, feature *m/z* and retention times are known, allowing selective MS/MS targeting of those ions within a certain chromatographic time range (allowing for minor retention differences). Before beginning targeted MS/MS experiments, we ensure the retention times of bradykinin, angiotensin, and neurotensin QA/QC peaks are comparable to those obtained during the dMS experiment. Simple linear adjustments in retention time can be applied to compensate for any retention time differences.

36. For high-resolution (i.e., FTMS/MS) ETD fragmentation, we used the following instrument parameters: 5 *m/z* isolation width, 15–20 ms reaction times, 1 μscan, 1E5 MS/MS precursor target value, 1E5 anion target value, and 1,200 ms max ion injection time. Reaction times are inversely related to precursor

charge state, and are therefore reduced for higher charges and increased for lower charge states. Current software allows for real time, charge state-dependent reaction time determination. For high-resolution (i.e., FTMS/MS) CID fragmentation, we used the following instrument parameters: 5 *m/z* isolation width, 30 ms reaction time, 35 normalized collision energy, 0.25 q-values, 1 μscan, 1E6 MS/MS precursor target value, and 1 s max ion injection time.

37. We prefer to use the accurate monoisotopic mass/charge and charge state of the precursor protein as the file name (e.g., 456.789_5_ETD.raw for the ETD spectra of target ion at *m/z* 456.789 with a charge of +5) because this value will later be required for database searching via ProSightPC v2.0.

38. The first search should be in Absolute Mass mode, which searches the database for full-length protein sequences which may or may not be modified. We use the following de-isotoping criteria: Minimum S/N = 3, Minimum RL = 0.9, Maximum Mass = 60,000, Maximum Charge = 25, First *m/z* = 300, and last *m/z* = 2,000. Input the appropriate monoisotopic precursor mass/charge and charge state. We use the following search criteria for first round searching: Search type = Absolute mass, Database Description = human, Precursor mass type = monoisotopic, Precursor Mass Tolerance = 2,000 Da, Fragment Mass Type = monoisotopic, Fragment Tolerance = 2.2 Da, Δm Mode = on, Disulfide = on, Include Modified Forms = on, # Min Matching Fragments = 4, PTM Handling = All PTMs. We consider all search scores of *P*-score <1E−5 and E-score <1E−5 as authentic protein matches.

39. We found that using the Sequence Gazer functionality of ProSightPC v2.0 allows for faster manual inspection of the search results.

40. The second search should be performed in Biomarker mode, which searches for protein sequences shorter than full-length proteins (e.g., truncated proteins via proteolysis). We use the following search criteria for second round searching: Search type = Biomarker Mode, Database Description = human, Precursor mass type = monoisotopic, Precursor Mass Tolerance = 2.2 Da, Fragment Mass Type = monoisotopic, Fragment Tolerance = 25 ppm, Δm Mode = on, Disulfide = on, Include Modified Forms = on, # Min Matching Fragments = 4, and PTM Handling = All PTMs. We consider all search scores of P-score <1e-5 and E-score <1e-5 as authentic protein matches.

41. Search criteria can be manually adjusted in attempts to identify these proteins. For example, search parameters can be relaxed (e.g., increase precursor and/or fragment mass tolerances,

reduce number of matching fragments) to increase the possibility of a protein match. Additionally, if a sequence tag can be identified, ProSightPC's Sequence Tag search mode can be used for protein identification. Lastly, manual interpretation of the spectrum may be required by an MS/MS subject matter expert.

42. The High-Throughput Wizard of ProSightPC v2.0 allows for the batch de-isotoping and database searching with search tree logic. We use the following de-isotoping parameters: Thrash Process Algorithm, with custom parameters Fragmentation MSn Analysis Level = ms2, Analyze only FTMS Fragmentation; Precursor Selection Options, Minimum S/N = 7, Maximum Charge = 30, Minimum Fit = 40, Remainder Threshold = 20, Precursor Selection Criterion = Highest Intensity, Allow Multiple Precursors with Relative Precursor Threshold = 10 %; and Fragmentation Analysis Options, Minimum S/N = 3.0, Maximum Charge = 30, Minimum Fit = 10, Remainder Threshold = 10, Minimum Fragmentation Base Peak Intensity = 100, with Add Remainder Afterwards. For the decision tree searching, we use search criteria as described in **Notes 34** and **36**. Briefly, first search by Absolute mass mode, Pass/Fail criteria equals *P*-score and *E*-score values of <1E–5, and second search Biomarker mode, Pass/Fail criteria equals *P*-score and *E*-score values of <1E–5.

43. Elucidator provides an annotation function that allows for saving of pertinent protein information, including sequence, protein name, and identity. If Excel is used, capture this information manually according to respective feature ID.

Acknowledgments

The authors thank Ekaterina Deyanova and Richard Seipert for their careful review of this manuscript. The authors also thank Ronald Hendrickson, Nathan Yates, and Daniel Spellman for their support of this work.

References

1. Blackstock WP, Weir MP (1999) Proteomics: quantitative and physical mapping of cellular proteins. Trends Biotechnol 17:121–127
2. Wilkins MR, Pasquali C, Appel RD, Ou K, Golaz O, Sanchez JC, Yan JX, Gooley A, Hughes G, Humphery-Smith I, Williams KL, Hochstrasser DF (1996) From proteins to proteomes: large scale protein identification by two-dimensional electrophoresis and arnino acid analysis. Biotechnology (N Y) 14:61–65
3. Aebersold R, Mann M (2003) Mass spectrometry-based proteomics. Nature 422:198–207
4. Ross PL, Huang YN, Marchese JN, Williamson B, Parker K, Hattan S, Khainovski N, Pillai S, Dey S, Daniels S, Purkayastha S, Juhasz P, Martin S, Bartlet-Jones M, He F, Jacobson A, Pappin DJ (2004) Multiplexed protein quantitation in saccharomyces cerevisiae using amine-reactive isobaric tagging reagents. Mol Cell Proteomics 3:1154–1169

5. Han H, Pappin DJ, Ross PL, McLuckey SA (2008) Electron transfer dissociation of iTRAQ labeled peptide ions. J Proteome Res 7:3643–3648
6. Ong SE, Blagoev B, Kratchmarova I, Kristensen DB, Steen H, Pandey A, Mann M (2002) Stable isotope labeling by amino acids in cell culture, SILAC, as a simple and accurate approach to expression proteomics. Mol Cell Proteomics 1:376–386
7. Ong SE, Kratchmarova I, Mann M (2002) Properties of 13C-substituted arginine in stable isotope labeling by amino acids in cell culture (SILAC). J Proteome Res 2:173–181
8. Lee AYH, Paweletz CP, Pollock RM, Settlage RE, Cruz JC, Secrist JP, Miller TA, Stanton MG, Kral AM, Ozerova NDS, Meng F, Yates NA, Richon V, Hendrickson RC (2008) Quantitative analysis of histone deacetylase-1 selective histone modifications by differential mass spectrometry. J Proteome Res 7:5177–5186
9. Liu H, Sadygov RG, Yates JR (2004) A model for random sampling and estimation of relative protein abundance in shotgun proteomics. Anal Chem 76:4193–4201
10. Meng F, Wiener MC, Sachs JR, Burns C, Verma P, Paweletz CP, Mazur MT, Deyanova EG, Yates NA, Hendrickson RC (2007) Quantitative analysis of complex peptide mixtures using FTMS and differential mass spectrometry. J Am Soc Mass Spectrom 18:226–233
11. Washburn MP, Wolters D, Yates JR (2001) Large-scale analysis of the yeast proteome by multidimensional protein identification technology. Nat Biotechnol 19:242–247
12. Wiener MC, Sachs JR, Deyanova EG, Yates NA (2004) Differential mass spectrometry: a label-free LC-MS method for finding significant differences in complex peptide and protein mixtures. Anal Chem 76:6085–6096
13. Zhao X, Deyanova EG, Lubbers LS, Zafian P, Li JJ, Liaw A, Song Q, Du Y, Settlage RE, Hickey GJ, Yates NA, Hendrickson RC (2008) Differential mass spectrometry of rat plasma reveals proteins that are responsive to 17+¦-estradiol and a selective estrogen receptor modulator PPT. J Proteome Res 7:4373–4383
14. Mazur MT, Cardasis HL, Spellman DS, Liaw A, Yates NA, Hendrickson RC (2010) Quantitative analysis of intact apolipoproteins in human HDL by top-down differential mass spectrometry. Proc Natl Acad Sci USA 107:7728–7733
15. Kelleher NL, Zubarev RA, Bush K, Furie B, Furie BC, McLafferty FW, Walsh CT (1999) Localization of labile posttranslational modifications by electron capture dissociation: the case of g-carboxyglutamic acid. Anal Chem 71:4250–4253
16. Zubarev RA, Horn DM, Fridriksson EK, Kelleher NL, Kruger NA, Lewis MA, Carpenter BK, McLafferty FW (2000) Electron capture dissociation for structural characterization of multiply charged protein cations. Anal Chem 72:563–573
17. Malek R, Metelmann-Strupat W, Zeller M, Muenster H (2005) Electron capture dissociation on the Finnigan LTQ FT-preserving posttranslational modifications during peptide fragmentation. Thermo Electron Application Note: 30081
18. McAlister GC, Phanstiel D, Good DM, Berggren WT, Coon JJ (2007) Implementation of electron-transfer dissociation on a hybrid linear ion trap-orbitrap mass spectrometer. Anal Chem 79:3525–3534
19. LeDuc RD, Taylor GK, Kim YB, Januszyk TE, Bynum LH, Sola JV, Garavelli JS, Kelleher NL (2004) ProSight PTM: an integrated environment for protein identification and characterization by top-down mass spectrometry. Nucleic Acids Res 32:W340–W345
20. Zamdborg L, LeDuc RD, Glowacz KJ, Kim YB, Viswanathan V, Spaulding IT, Early BP, Bluhm EJ, Babai S, Kelleher NL (2007) ProSight PTM 2.0: improved protein identification and characterization for top down mass spectrometry. Nucleic Acids Res 35:W701–W706
21. Joy T, Hegele RA (2008) Is raising HDL a futile strategy for atheroprotection? Nat Rev Drug Discov 7:143–155
22. Vaisar T, Pennathur S, Green PS, Gharib SA, Hoofnagle AN, Cheung MC, Byun J, Vuletic S, Kassim S, Singh P, Chea H, Knopp RH, Brunzell J, Geary R, Chait A, Zhao XQ, Elkon K, Marcovina S, Ridker P, Oram JF, Heinecke JW (2007) Shotgun proteomics implicates protease inhibition and complement activation in the antiinflammatory properties of HDL. J Clin Invest 117:746–756
23. Gordon SM, Deng J, Lu LJ, Davidson WS (2010) Proteomic characterization of human plasma high density lipoprotein fractionated by gel filtration chromatography. J Proteome Res 9:5239–5249
24. Heinecke JW (2009) The HDL proteome: a marker-and perhaps mediator-of coronary artery disease. J Lipid Res 50:S167–S171
25. Rezaee F, Casetta B, Levels J, Speijer D, Meijers J (2006) Proteomic analysis of high-density lipoprotein. Proteomics 6:721–730
26. Chapman MJ, Goldstein S, Lagrange D, Laplaud PM (1981) A density gradient ultracentrifugal

procedure for the isolation of the major lipoprotein classes from human serum. J Lipid Res 22:339–358

27. Liu P, O'Mara B, Warrack B, Wu W, Huang Y, Zhang Y, Zhao R, Lin M, Ackerman M, Hocknell P, Chen G, Tao L, Rieble S, Wang J, Wang-Iverson D, Tymiak A, Grace M, Russell R (2010) A tris (2-carboxyethyl) phosphine (TCEP) related cleavage on cysteine-containing proteins. J Am Soc Mass Spectrom 21: 837–844

28. Mueller LN, Brusniak MY, Mani DR, Aebersold R (2008) An assessment of software solutions for the analysis of mass spectrometry based quantitative proteomics data. J Proteome Res 7:51–61

Chapter 11

Quantitative Proteomics Analysis of High-Density Lipoproteins by Stable ^{18}O-Isotope Labeling

Elena Burillo, Jesus Vazquez, and Inmaculada Jorge

Abstract

For the large-scale study of dynamic proteomes, quantitative proteomic approaches based on stable isotope labeling and mass spectrometry (MS) have been developed as a high-throughput, reproducible, and robust alternative to conventional gel-based techniques. In this chapter, we describe in detail a quantitative proteomic strategy based on HDL isolation by affinity chromatography, in-gel trypsin digestion of protein extracts, peptide ^{18}O labeling, separation by off-gel isoelectric focusing, and peptide analysis on a linear ion trap mass spectrometer, followed by the application of a robust multilayered statistical model. This protocol is of universal applicability and has been successfully applied to the global characterization of the HDL proteome with some specific considerations for this particle, paving the way to the in-depth study of the protein cargo of HDL and its implication in cardiovascular diseases.

Key words Mass spectrometry, Quantitative proteomics, High-density lipoprotein, ^{18}O stable isotopic labeling

1 Introduction

Many observational prospective studies have confirmed the protective role of high-density lipoprotein (HDL) cholesterol in cardiovascular disease (CVD) (1). The best-studied protective function of HDL is the reverse cholesterol transport from peripheral tissues to the liver (2). However, it is now so well-known that HDL particles have other interesting functions related to innate immunity, endothelial vascular function, protease activity regulation, oxidation, thrombosis, and inflammation (3). Therefore, in the last years, the interest in HDL proteome has raised due to the fact that the functional status of HDL, which is largely dependent on its protein components, is probably an important determinant of CVD (4).

Mass spectrometry (MS)-based quantitative proteomics plays an increasingly important role in cardiovascular research. Specifically,

Fernando Vivanco (ed.), *Vascular Proteomics: Methods and Protocols*, Methods in Molecular Biology, vol. 1000, DOI 10.1007/978-1-62703-405-0_11,

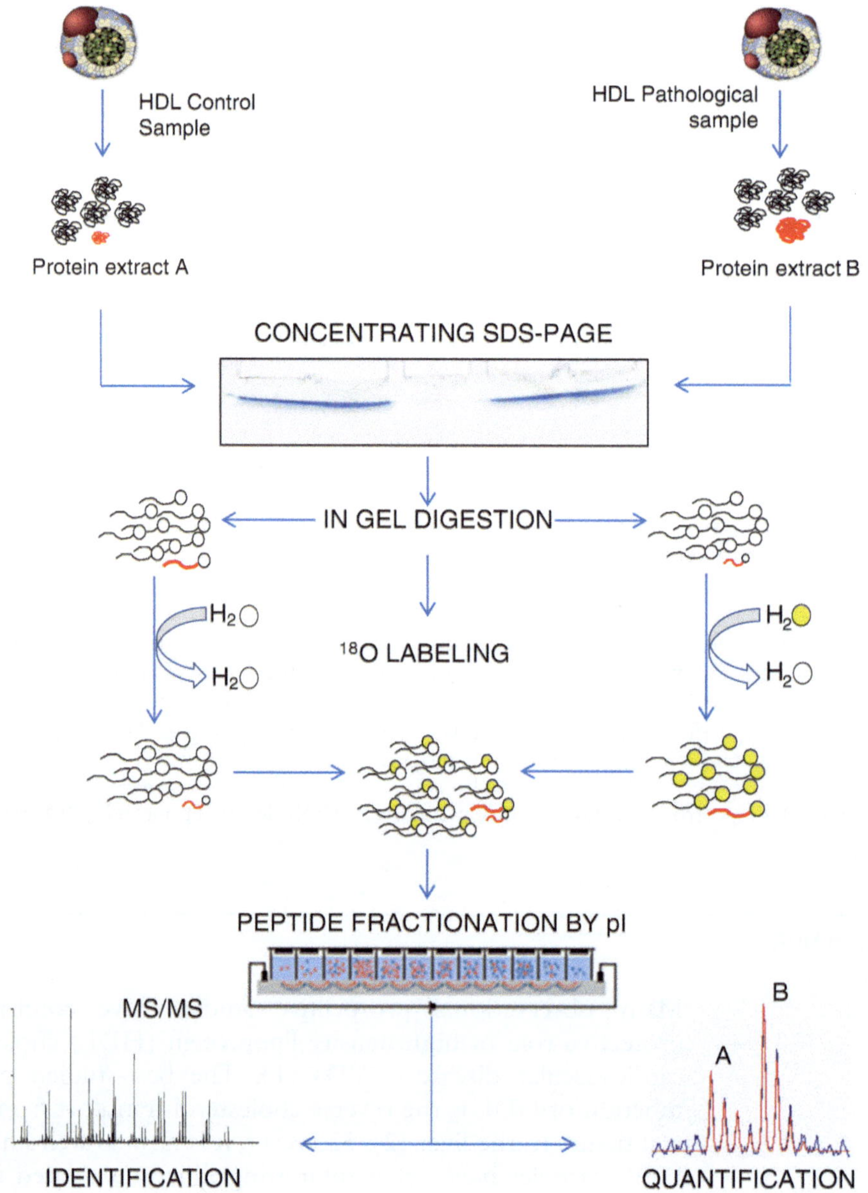

Fig. 1 Workflow for the proposed ^{18}O-based quantification protocol of the HDL proteome

^{18}O labeling is a powerful tool mainly due to the simplicity of the enzymatic reaction, the absence of side reactions, the possibility of controlling the labeling efficiency, and its general applicability to biological models that cannot be metabolically labeled. In this protocol we present a standardized high-throughput quantitative proteomics method based on whole proteome concentration by SDS-PAGE, optimized in-gel digestion, peptide ^{18}O labeling, and separation by off-gel isoelectric focusing followed by liquid chromatography coupled to an electrospray–linear ion trap (LC–ESI–LIT) mass spectrometer (5) (Fig. 1).

Our protocol provides a general approach to perform quantitative proteomics by ^{18}O labeling in high-throughput studies, which allows controlling for the biological variability associated to HDL proteome isolated from plasma.

2 Materials

2.1 Equipment

1. Centrifuge and minicentrifuges.
2. FPLC system (ÄKTA Purifier UPC 10, GE Healthcare).
3. Amicon Ultra 10 KDa filters (Millipore).
4. Pipettors and multichannel pipettors.
5. Water bath up to 100 °C.
6. UV–VIS spectrophotometer.
7. Mini Protean Tetra Cell for SDS-PAGE (BioRad).
8. Power supply.
9. NHS-activated Sepharose™ High Performance 5 mL (GE Healthcare).
10. C18 cartridge (Oasis and OMIX cartridge, Waters).
11. Orbital shaker.
12. Heater.
13. pH indicator paper strips.
14. Parafilm.
15. Vacuum centrifuge dryer.
16. Wizard minicolumns (Promega).
17. Bath sonicator.
18. Vortex.
19. 3100 OFFGEL Fractionator (Agilent).
20. ImmobilineTM DryStrip Gels pH 3–10, 24 cm (GE Healthcare).
21. Linear ion trap mass spectrometer (LTQ, Thermo Scientific).
22. Surveyor HPLC system.

2.2 Buffers and Reagents

1. FPLC affinity isolation of HDL:
   - (a) Pefabloc SC (Roche).
   - (b) Standard coupling solution: 0.2 M $NaHCO_3$, 0.5 M NaCl, pH 8.3.
   - (c) Anti-ApoAI antibody (BioDesign International).
   - (d) Buffer A: 0.5 M ethanolamine, 0.5 M NaCl, pH 8.3.
   - (e) Buffer B: 0.1 M sodium acetate, 0.5 M NaCl, pH 4.

(f) Washing buffer: 0.1 M NaHCO3, 0.5 M NaCl, 1 mM EDTA, pH 8.0.

(g) Elution buffer: 0.1 M glycine, 0.5 M NaCl, 10 % (v/v) dioxane, pH 2.8.

(h) Equilibration buffer: 0.5 M Tris–HCl, pH 8.0.

2. Concentrating gel electrophoresis and in-gel digestion:

(a) Sample buffer 4×: 5 % (w/v) SDS, 10 % (v/v) glycerol, 25 mm Tris–HCl, pH 6.8, 10 mM dithiothreitol (DTT), and 0.01 % (w/v) bromophenol blue.

(b) 10 % (w/v) ammonium persulfate (APS).

(c) *N,N,N,N*-tetramethylethylenediamine (TEMED) .

(d) 1.5 M Tris–HCl pH 8.8.

(e) 0.5 M Tris–HCl pH6.8.

(f) Acrylamide/bisacrylamide solution 30 % (37.5:1) (BioRad).

(g) 10 % (w/v) sodium dodecyl sulfate (SDS).

(h) Running buffer: 0.025 M Tris pH 8.3, 0.192 M glycine, and 0.1 % (w/v) SDS.

(i) GelCode Blue Stain Reagent (Thermo Scientific).

(j) Fixing solution: 40 % (v/v) methanol / 10 % (v/v) acetic acid.

(k) Digestion buffer: 50 mM ammonium bicarbonate pH 8.8, 10 % (v/v) acetonitrile (ACN).

(l) Trypsin sequencing grade (Promega).

3. ^{18}O labeling:

(a) Immobilized TPCK trypsin (Thermo).

(b) 0.5 M sodium citrate pH 6.0.

(c) $H_2{}^{18}O$ (Sigma-Aldrich).

(d) Serine protease inhibitor TLCK (Sigma-Aldrich).

4. Peptide fractionation by isoelectric point with OFFGEL Fractionator Kit:

(a) Focusing buffer: 5 % (v/w) glycerol, IPG buffer pH 3–10 (dilute 1:50) (GE Healthcare), in HPLC grade water.

(b) Mineral oil.

5. Mass spectrometric analysis by LC–ESI–LIT:

(a) Loading buffer for ^{18}O-labeled peptides: 5 mM ammonium formate pH 3.0.

(b) Reagents for mass spectrometry should be HPLC grade: solvent A, 0.1 % (v/v) formic acid; solvent B, 0.1 % (v/v) formic acid, 80 % (v/v) ACN.

(c) Calibration solution (always prepare fresh): 20 μL of caffeine solution stock (1 mg/mL in methanol, Sigma), 0.4 μL of MRFA solution stock (3 mg/mL in 50 % (v/v) methanol, Thermo Finnigan), 30 μL of Ultramark stock (10 μL of Ultramark in 10 μL of acetonitrile, Alfa Aesar), and 0.4 μL of formic acid, and 19.8 μL of water.

(d) Tune solution (bovine serum albumin, Michrom): prepare 200 μL in 25 % (v/v) acetonitrile and 0.1 % (v/v) formic acid at a concentration of 50 fmol/μL.

6. Common reagents used through the protocol:
   (a) HPLC grade water.
   (b) HCl.
   (c) PBS pH 7.4.
   (d) Bradford protein assay reagent (BioRad).
   (e) Bovine serum albumin.
   (f) HPLC grade methanol.
   (g) Acetic acid.
   (h) HPLC grade acetonitrile (ACN).
   (i) Dithiothreitol (DTT).
   (j) Iodoacetamide.
   (k) Ammonium bicarbonate.
   (l) Trifluoroacetic acid (TFA) 25 %.
   (m) Ammonium acetate.
   (n) Formic acid.
   (o) Isopropanol.

3 Methods

3.1 FPLC Affinity Isolation of HDL

There are different types of HDL isolation methods. In this protocol, the selected method is based on immunoaffinity against ApoAI, which is the main protein present in HDL. This method allows not only a fast isolation of particles containing ApoAI or LpAI but also a concomitant depletion of albumin and other high-abundance plasma proteins.

The first step is to bind the antibody against ApoAI to the resin. We use a 5 mL NHS-activated Sepharose™ High Performance prepacked column (GE Healthcare) for ligand coupling via primary amines. Since this column operates at a 0.3 MPa, it is ideally fitted to an FPLC system. Alternatively, the column could be used with syringe pumping, but it is highly recommendable to use an automated system for optimal reproducibility.

1. Plasma samples:
   - (a) Collect blood samples from fasted (≥12 h) subjects. Obtain 10 mL of blood with EDTA as anticoagulant.
   - (b) Centrifuge blood samples at 3,000 × *g* for 15 min at 4 °C.
   - (c) In order to prevent proteolysis, add the protease inhibitor Pefabloc SC (Roche) at a concentration of 0.5 mM.
   - (d) Plasma samples can be stored at −80 °C until further use.
2. Column preparation:
   - (a) Follow strictly the NHS-activated Sepharose™ High Performance column instructions provided by the manufacturer.
   - (b) The buffers used must be of HPLC grade purity. It is also strongly recommended to filter the buffers by passing them through a 0.45 μm filter before use.
   - (c) Prepare the standard coupling solution: 0.2 M $NaHCO_3$, 0.5 M NaCl, pH 8.3.
   - (d) Dissolve the antibody against ApoAI in the standard coupling solution to achieve a final concentration in the 0.5–10 mg/mL range in a final volume of 5 mL.
   - (e) Bind the antibody to the column following the instructions specified by the manufacturer. First, remove the top-cap and apply a drop of ice cold 1 mM HCl to the top of the column to avoid air bubbles. Then, connect the adequate connector to the top of the column and remove the snap-off end at the column outlet.
   - (f) Add 3× 10 mL of 1 mM HCl at a maximum flow of 5 mL/min to wash out the column of isopropanol (*see* **Note 1**).
   - (g) Add immediately the anti-ApoAI antibody and seal the column. The binding reaction is performed during 30 min at room temperature (*see* **Note 2**).
   - (h) After antibody binding, deactivation and washing of the column are key steps. Prepare Buffer A (0.5 M ethanolamine, 0.5 M NaCl, pH 8.3) and Buffer B (0.1 M sodium acetate, 0.5 M NaCl, pH 4).
   - (i) Inject 3× 10 mL of Buffer A; then, 3× 10 mL of Buffer B, and, finally, 3× 10 mL of Buffer A.
   - (j) Leave the column for 30 min at room temperature.
   - (k) Inject 3× 10 mL of Buffer B; then, 3× 10 mL of Buffer A, and, 3× 10 mL of Buffer B.
   - (l) To neutralize column pH, add 10 mL of a neutral buffer, for example, phosphate buffer saline (PBS) pH 7.4 (*see* **Note 3**).

3. HDL isolation:
   (a) First, prepare the necessary buffers. Washing buffer: 0.1 M $NaHCO_3$, 0.5 M NaCl, 1 mM EDTA, and pH 8.0. Elution buffer: 0.1 M glycine, 0.5 M NaCl, 10 % (v/v) dioxane, and pH 2.8. Equilibration buffer: 0.5 M Tris–HCl, pH 8.0.
   (b) Inject 5 mL of plasma in the FPLC system coupled to the anti-ApoAI affinity column.
   (c) Use 10 mL of PBS to ensure the equilibration of the column.
   (d) Perform the binding reaction using the washing buffer at a flow rate of 0.5 mL/min with a maximum pressure of 0.3 MPa.
   (e) When the nonbinding fraction is washed (*see* **Note 4**), change buffer to elution buffer at a flow rate of 1.5 mL/min (*see* **Note 5**).
   (f) Collect the binding fraction (HDL particles) as a single 5 mL fraction over 1 mL of equilibration buffer to quickly neutralize the acidic pH and to prevent protein degradation.
   (g) Concentrate the HDL purified fraction to 200 μL using Amicon Ultra 10 kDa filters (Millipore) at 4,000 g in a swinging bucket rotor or at 5000 g in a fixed angle rotor for 35–40 min at 4 °C.
   (h) Calculate total protein concentration by Bradford (BioRad), using BSA (bovine serum albumin) as a standard.

3.2 Concentrating Gel Electrophoresis and In-Gel Digestion

This protocol details the instructions to concentrate the HDL proteome in a single protein band and to perform in-gel digestion.

1. Electrophoresis:
   (a) Suspend 100 μg of HDL in the appropriate volume of sample buffer 4× (5 % (w/v) SDS, 10 % (v/v) glycerol, 25 mM Tris–HCl, pH 6.8, 10 mM dithiothreitol (DTT), and 0.01 % (w/v) bromophenol blue).
   (b) Heat the samples during 5 min at 95 °C.
   (c) Prepare a conventional SDS-PAGE gel (1.5 mm thick, 4 % stacking, and 10 % resolving).
   (d) Apply the samples and run the electrophoresis at 15 V until the front enters 3 mm into the resolving gel (*see* **Note 6**).
   (e) Fix the proteins with fixing solution (methanol 40% (v/v)/ acetic acid 10 % (v/v)) during 30 min at room temperature.
   (f) Stain the single protein bands by GelCode Blue Stain Reagent (Thermo Scientific) at room temperature overnight.
   (g) Excise the protein bands horizontally and cut them into pieces (2 × 2 mm). Pool the gel pieces in a tube (one tube per sample).

2. Reduction and alkylation of proteins (*see* **Note 7**):
   (a) Wash the gel pieces twice with 1 mL of HPLC grade water during 10 min in continuous shaking.
   (b) Remove the supernatant (SPN) and add 1 mL of acetonitrile (ACN) 100 % to dehydrate the gel pieces. Shake during 10 min and repeat again.
   (c) Remove the SPN and add 800 μL of 10 mM DTT in 25 mM ammonium bicarbonate, pH 8.8. Shake for 1 h.
   (d) Remove the SPN and dehydrate the gel pieces by shaking with 800 μL of ACN 100 % during 10 min.
   (e) Remove the SPN and shake the gel pieces during 1 h in the darkness with 800 μL of 54 mM iodoacetamide in 25 mM ammonium bicarbonate, pH 8.8.
   (f) Remove the SPN and add 800 μL of ACN 100 %. Shake for 10 min.
   (g) Remove SPN and add 500 μL of 50 mM ammonium bicarbonate pH 8.8. Shake for 15 min.
   (h) Without removing the SPN, add 500 μL of ACN 100 %. Shake for 15 min.
   (i) Remove the SPN and dry down the gel pieces in a vacuum centrifuge.
3. Digestion:
   (a) Prepare the digestion buffer: ammonium bicarbonate 50 mM pH 8.8, 10 % (v/v) ACN.
   (b) Prepare 20 μg of trypsin in a 1:5 (w/w) enzyme to protein ratio. Add digestion buffer to achieve a final concentration of 80 ng /μL.
   (c) Add to each sample 250 μL of trypsin and incubate 1 h on ice. This allows the enzyme to soak into the gel pieces while the enzymatic activity remains inhibited.
   (d) Add digestion buffer to a final trypsin concentration of 60 ng/μL. Check that pH is near 8.0.
   (e) Seal the tubes with Parafilm and incubate overnight with vigorous shaking at 37 °C.
4. Peptide extraction and desalting (*see* **Note 7**):
   (a) After digestion, leave the tubes at room temperature for 15 min.
   (b) Centrifuge the tubes at 15,600 × *g* during 30 s.
   (c) Collect the SPN in a 2 mL clean tube.
   (d) Add 500 μl of 12 mM ammonium bicarbonate pH 8.8, and incubate for 1 h.

(e) Collect the SPN and add 25 % (v/v) trifluoroacetic acid (TFA) for a 1 % final concentration. Vortex the mixture and centrifuge at maximum speed during 15 s (spin down).

(f) Check the pH, it should be <3.

(g) Desalt each sample with a C18 reversed phase OMIX cartridge (Waters) (*see* **Note 8**). Wash the column adding 50 μL of 50 % (v/v) ACN (×2) followed by 50 μL of 100 % ACN (×2). Equilibrate the column with 50 μL of 0.1 % (v/v) TFA. Repeat the equilibration step three times. Once the column is equilibrated, load and elute the sample. Collect also the pass-through fraction. Wash again with 100 μL of 0.1 % (v/v) TFA. Repeat. Elute twice with 50 μL of 0.1 % (v/v) TFA/50 % (v/v) ACN. In the second elution, flush out the mobile phase to ensure complete peptide recovery.

(h) Dry down the elution fraction in a vacuum centrifuge.

3.3 ^{18}O Labeling

This isotopic labeling, based on a simple enzymatic reaction, facilitates the quantitation of paired samples (e.g., diseased vs. control). This work illustrates the application of ^{18}O labeling to HDL samples.

1. Immobilized trypsin preparation:
   (a) Prepare the beads containing immobilized trypsin keeping a 1:5 (w/v) protein: enzyme ratio.
   (b) Shake the bead suspension vigorously.
   (c) Add 80 μL of beads to 1.5 mL clean tubes (one tube for each sample).
   (d) Clean the beads by centrifugation at 13,000 rpm during 1 min with 320 μL of 500 mM ammonium acetate pH 6. Repeat the cleaning step six times.
   (e) Remove the SPN and dry down the beads in a vacuum centrifuge.
2. Dry down a 150 μL aliquot of 0.5 M sodium citrate, pH 6 per sample. Sodium citrate is used to maintain the pH constant during labeling reaction.
3. Redissolve the sodium citrate in 100 μl of $H_2{}^{18}O$ (for the experimental sample) or $H_2{}^{16}O$ (for the control sample). It is essential to keep the following water/protein ratio: 1 mL of H_2O (^{18}O or ^{16}O) per mg of protein. Vortex vigorously and centrifuge the solutions.
4. Add the sodium citrate to the dried trypsin beads and check that pH is 6.

5. Resuspend the peptides in 25 μl of 100 % ACN for a final ACN concentration of 20 % (v/v) ACN.
6. Sonicate peptides during 5 min in a bath sonicator.
7. Add the immobilized trypsin beads (labeled and non-labeled aliquots) to the corresponding peptide samples.
8. Seal the tubes and incubate overnight at 37 °C. Shake at 300 rpm keeping the tubes in horizontal position to prevent the formation of trypsin pellets.
9. After overnight labeling, leave the tubes at room temperature for 15 min.
10. Centrifuge samples during 15 s at maximum speed (spin down).
11. Filter the beads with a Wizard minicolumn (Promega):
    (a) Clean the column by centrifugation (30 s at 13,000 rpm) with 200 μL of 100 % ACN.
    (b) Place the column on a clean tube.
    (c) Load column with sample and elute retentate by centrifugation (30 s at 13,000 rpm).
    (d) In order to maximize peptide recovery, centrifuge the column at 15,600 × *g* during 15 min.
12. To inactivate residual amounts of trypsin that may have been liberated from the beads, prepare 50 mg/mL TLCK in methanol. Add TLCK to the eluted samples for a 1 mM final concentration.
13. Vortex gently the sample inhibitor mixture and spin down.
14. Seal the tubes with Parafilm and incubate for 1 h at 37 °C (shake at 800 × *g*).
15. At this point, combine samples (the experimental and control samples to be compared) in the same tube.
16. Add 1 M ammonium formate pH 3.0 to a final 1.5 % (v/v) ACN concentration.
17. Desalt the peptide mixture with a C18 Oasis cartridge:
    (a) Wash the column adding 1 mL of 50 % (v/v) ACN followed by 50 μL of 100 % ACN (×2).
    (b) Equilibrate the column with 5 mM ammonium formate pH 3.0. Repeat the equilibration step three times.
    (c) Load the sample onto the equilibrated column. Elute the peptides. Collect also the pass-through fraction.
    (d) Wash twice with 100 μL of 5 mM ammonium formate pH 3.0.
18. Elute twice with 50 μL of 5 mM ammonium formate pH 3.0/50% (v/v) ACN. In the second elution, flush out the mobile phase for optimal peptide recovery.

3.4 Peptide Fractionation by Isoelectric Point (pI)

The protocol explained in the Agilent 3100 OFFGEL Fractionator Kit Quick Star guide should be strictly followed. Briefly:

1. Clean the tray and electrodes with isopropanol first and then with methanol. Afterwards dry with compressed air.
2. Prepare focusing buffer: 5 % (v/v) glycerol, IPG buffer pH 3–10 (dilute 1:50), in MilliQ water.
3. Place the IPG strip pH 3–10 and the 24 well frames on the tray.
4. Add 40 μL of focusing buffer into each well. Be careful not to touch the gel with the tip.
5. Using tweezers, wet paper pads (4 per IPG strip) with HPLC grade water and place two wetted pads at each end of the IPG strip. The upper pad should be replaced every 24 h of fractionation. Make sure that there is no gap between the pad and the frame.
6. Incubate 15 min.
7. Resuspend the proteome in 3.8 mL of focusing buffer and vortex during 15 s.
8. Load 150 μL of sample per well.
9. Finish the procedure by assembling the IPG strips, frames, and electrodes.
10. Run the fractionation method 24PE (10–72 h) detailed in the manufacturer's guide with minor changes: limit voltage to 4,000 V instead of 8,000 V to prevent excess heating of the strip (*see* **Note 9**).
11. Once the method is finished, collect the fractions in clean tubes, using one clean tube per well. Clean the tray and electrodes as indicated in **step 1**.
12. After off-gel fractionation, desalt peptides. A multichannel pipettor could be used. Add 20 μL of 1 M ammonium formate pH 3.0 to each recovered fraction. Check pH (it should be <3).
13. Desalt peptides using the OMIX columns:
    (a) Add twice 50 μL of 50 % (v/v) ACN.
    (b) Add twice 50 μL of 100 % ACN.
    (c) Add three times 50 μL of 5 mM ammonium formiate pH 3.0/3% (v/v) ACN. This residual amount of ACN is necessary in order to minimize binding of ampholytes to the OMIX columns.
    (d) Load the sample and collect the pass-through fraction.
    (e) Add twice 100 μL of 5 mM ammonium formate pH 3/3% (v/v) ACN.
    (f) Add twice 50 μL of 5 mM ammonium formate pH 3/50 % (v/v) ACN.
14. Dry down the elution fraction in a vacuum centrifuge.

3.5 Mass Spectrometric Analysis by LC–ESI–LIT

Peptide analysis in electrospray-based mass spectrometers is usually performed by HPLC separation coupled online to MS detection (6).

1. This method has been optimized to use on a Surveyor HPLC system coupled to a linear ion trap (LIT) mass spectrometer (model LTQ, Thermo Scientific).
2. Resuspend each tryptic peptide fraction in 20 μL of loading buffer, vortex during 1 min, and spin down.
3. Concentrate and desalt the peptides on an RP pre-column (0.32 × 30 mm BioBasic-18, Thermo Electron) and online elute onto an analytical RP column (0.18 × 150 mm BioBasic-18, Thermo Electron) operating at 1.5 μL/min. After 10 min equilibration with solvent A, elute peptides using the following gradient: 0–10 % B in 2 min, 10–40 % B in 120 min, 40–95 % in 2 min, and 95 % for 5 min.
4. The electrospray ion source is operated with 3.3 kV potential, using nitrogen gas to improve the spray, and a capillary temperature of 200 °C. The ion trap should be previously calibrated using a solution containing caffeine, MRFA, and Ultramark 1621 as recommended by the manufacturer (*see* **Note 10**). Tune the instrument using a commercial tryptic digest of bovine serum albumin, programming the enhanced scan type, and collecting the parent ions in profile mode (*see* **Note 11**). Use the manufacturer's guide.
5. Peptides identified in survey scans are subjected to a high-resolution scanning mode (ZoomScan), which is used to quantify the isotopic partners, and then subjected to MS/MS analysis. This allowed both identification and relative quantification of peptides. The LIT is operated in data-dependent ZoomScan and MS/MS switching mode, using the six most intense precursors detected in a survey scan from 400 to 1,600 *m/z* (7, 8). The maximum injection time is set to 200 ms for the parent ion scan and 100 ms for the ZoomScan and the MS/MS scan. The number of microscans is set to 3. Select a 12 Da window for ZoomScan to allow the monitoring of the entire $^{16}O/^{18}O$ isotopic envelope of doubly or triply charged peptides, regardless of the specific isotopic peak chosen as a precursor (singly charged ions are excluded), and select a 3 Da mass window to fragment selected parent ions. The normalized collision energy is set to 35 % and a dynamic exclusion is applied during 75 s periods and with an exclusion list size of 500, to avoid fragmenting each ion more than twice (9).

3.6 Protein Identification

The most widespread database searching algorithms is SEQUEST (Thermo Finnigan) (10). This software computes a cross correlation (X_{corr}) function to assess the quality of the match between a tandem mass spectrum and the amino acid sequence information present in a database, together with the parameter called ΔC_n, which is the normalized score obtained from the difference between the first- and the second-ranked sequences (*see* **Note 12**).

1. Search the MS/MS raw files acquired from the mass spectrometer against the human UniProt database (http://www.uniprot.org), using the SEQUEST algorithm.
2. Search the same collection of MS/MS spectra against an inverted database obtained from the above database.
3. Use the following parameters for the searches: tryptic cleavage after Arg and Lys, up to two missed cleavage sites, and tolerances of ±2 Da for precursor ions and of ±1.2 Da for MS/MS fragment ions. Include as variable modifications Met oxidation, and Lys and Arg +4 Da labeling. Consider carbamidomethylation of Cys as a fixed modification (11).
4. In large-scale experiments an appropriate indicator of significance is the False Discovery Rate (*FDR*), defined as the proportion of false positives among the population of spectra passing a given *p*-threshold (11, 12). A very accurate estimation of the *FDR* is performed by the Probability Ratio, a non-parametric, robust indicator based on spectral quality which takes into account the information provided by the best and the second best scores, together with the refined target–decoy search strategy (13–15).

3.7 Protein Quantification

Peptide labeling with ^{18}O tags is an enzymatic process which produces the incorporation of two ^{18}O atoms at the C terminus of peptides and, therefore, adds either 2- or 4-Da mass tags depending on the number of oxygen atoms exchanged (16). During the labeling reaction, two new labeled species are produced containing one (B_1) or two (B_2) ^{18}O atoms at the C terminus, in addition to the unlabeled chemical species (B_0) (17) (Fig. 2).

1. The relative quantification of a given peptide is calculated as $\log_2 (A/B)$, where A is the amount of non-labeled peptide originating from the ^{16}O-labeled sample and B the amount of peptide coming from the ^{18}O-labeled sample ($B = B_0 + B_1 + B_2$).
2. The $\log_2(A/B)$ ratios are calculated from medium-resolution ZoomScan spectra, which are fitted to a theoretical curve describing the expected isotopic envelope of the identified peptide, so that the best-fit parameters allow a simultaneous calculation of the peptide concentration in the two samples (A and B) and of the labeling efficiency f (*see* **Note 13**).
3. The proportion of non-labeled (B_0) and mono- (B_1) and di-labeled (B_2) peptides in the labeled sample (B) can be determined from these parameters (*see* **Note 14**). The theoretical curve used to fit the isotopic peaks is a mixed Gaussian/double exponential distribution (*see* **Notes 15** and **16**).

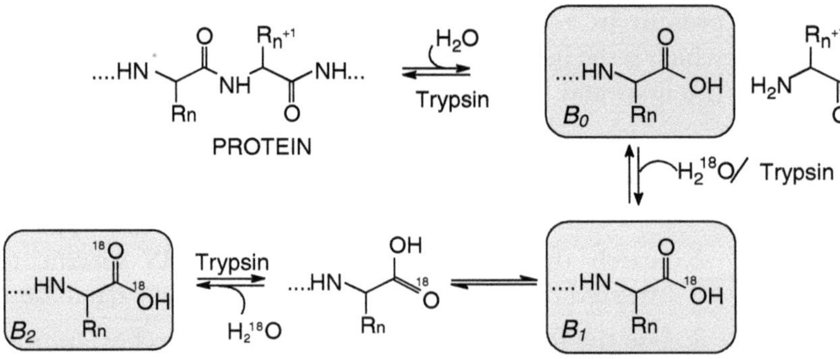

Fig. 2 ^{18}O-labeling reaction. In the first step, the peptide bond is cleaved by a reaction catalyzed by trypsin in the presence of $H_2{}^{16}O$, generating the chemical species B_0, with two ^{16}O atoms at the peptide C terminus. In the second step, trypsin allows the transfer of one or two ^{18}O atoms to the C terminus, generating two new chemical species: B_1, with one atom of ^{18}O, and B_2, with two atoms of ^{18}O

3.8 Statistical Analysis

Statistical analysis of quantitative data was performed with a novel random-effects model that assumes local normality of the measures and variance homogeneity (*see* **Note 17**) and includes four different sources of error: at the spectrum-fitting, scan, peptide, and protein levels (18).

1. Using this model, the $\log_2$-ratio of peptide concentration in samples *A* (non-labeled) and *B* (labeled) determined by scan s coming from peptide p derived from protein q is expressed as $X_{qps} = \log_2 (A/B)$. When the peptide amounts in the two samples (*A* and *B*) are identical, $X_{qps} = 0$. Experimental deviations from this value are assumed to come from a systematic error in the ratio in which the two samples are mixed, μ; from deviations of protein concentration due to biological variability and errors committed during the process of preparation of the protein extracts, ρ_q; from deviations in peptide concentration due to peptide preparation from the protein extracts, β_{qp}; and from errors committed during the quantification of peptide pair from the scan, ξ_{qps} (Fig. 3).
2. Quantification data is performed taking into account that not all spectra produce equally precise quantifications. Each spectrum has a different variance, and hence, data cannot be treated as a whole by using a unique normal distribution. Thus, the statistical weight associated to the scan, Wqps, is calculated from the spectrum-fitting and the scan variance, σ_s^2, as described (18).
3. The $\log_2$-ratio value associated to each peptide, X_{qp}, is calculated as a weighted average of the scans used to quantify the peptide, and the value associated to each protein, X_q, is similarly the weighted average of its peptides. Besides, a grand mean, *X*, is calculated as a weighted average of the protein values. In turn, the statistical weight associated to each peptide,

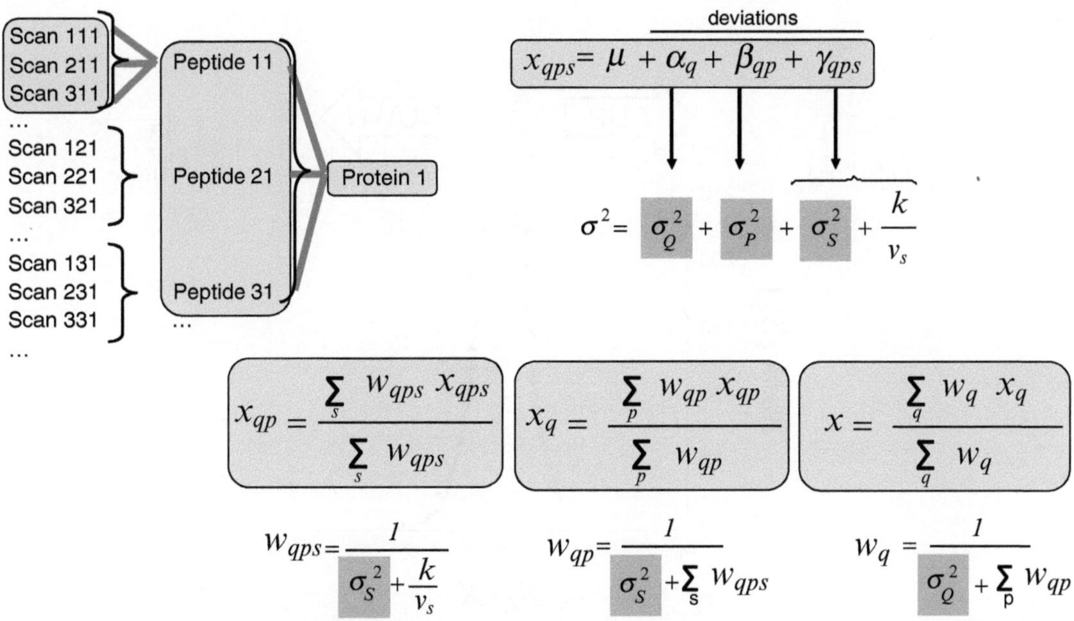

Fig. 3 Statistical model for quantitative proteomics. At each level (scan, peptide, and protein), the ¹⁸O quantification is carried out with weighted averaged measurements for the next lower level, so that more precise measurements contribute to a greater extent to the final average. The final error of the measurement for a protein is made up of the sum of the errors for each level of measurement. Details about the statistical model and the algorithm used to calculate the variances at the scan, peptide, and protein levels can be found in (18)

W_{qp}, is calculated from the corresponding scan weights and the peptide variance, σ^2_p, and that of each protein, W_q, is calculated from the corresponding peptide weights and the protein variance, σ^2_q. In all cases, the statistical weights are the inverses of variances (Fig. 3).

4. Outliers at the scan and peptide levels are detected by calculating the probability that the measurements deviate from the expected average according to their respective variances and controlling for the *FDR* at each level, FDR_{qps}, and FDR_{qp}, respectively. Scan outliers are usually explained by the presence of other isobaric peptides or contaminants and due to a poor curve fitting. Peptide outliers are explained due to partial protein digestion and methionine oxidation. These outliers should be checking, visually inspected and eliminated from the analysis.
5. Significant protein expression changes are outliers at protein level and are detected by using the same strategy as that used at the scan and peptide level. Determinations having an FDR_q lower than 5 % were considered significant changes in abundance, which correspond to protein average higher than that expected from grand mean (Fig. 4).

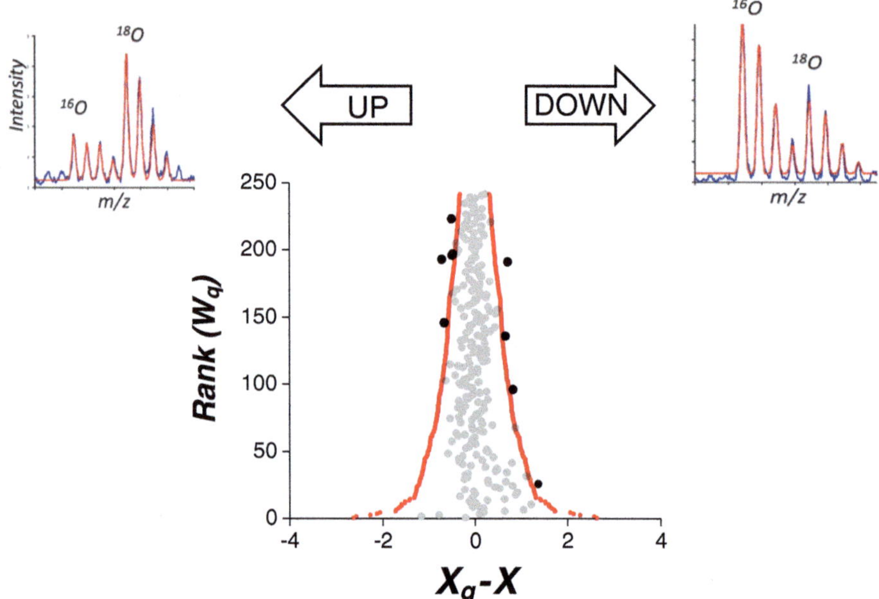

Fig. 4 Analysis of statistically significant protein expression changes. Example of a large-scale ^{18}O-labeling experiment of on the proteome of HDL particle. Data are represented as the weight distribution of protein quantifications around their corresponding grand means; *black points* indicate significant increases (*towards the left*) or decreases (*towards the right*) in protein abundance at 5 % FDR criterion (*red lines*)

4 Notes

1. At this point, be careful not to use excess flow, as this could result in a compressed medium.
2. Binding could be also performed during 4 h at 4 °C.
3. If the column is not used immediately, then it should be stored in storage solution: 0.05 M Na_2HPO_4, 0.1 % (w/v) NaN3, pH 7.0.
4. The nonbinding fraction is easily detected as a very intense peak (aromatic amino acids absorb at 280-nm wavelength).
5. This step should be carried out as quickly as possible to improve HDL release and to prevent column damage due to acidic pH.
6. To avoid the gel cutting effect, voltage was stopped when the front of the run entered 3 mm into the resolving part of the gel, so that all proteins got inside the stacking gel but remained unseparated forming a single band around the stacking and resolving gel interface. The bands from each sample were then cut into pieces and subjected to in-gel trypsin digestion in one single tube per sample, allowing a maximum reproducibility due to all the proteins were digested together.

7. It is recommended to shake the tubes horizontally at 300 rpm in an orbital shaker.
8. During the washing, equilibration, and sample elution steps, it is important to prevent the column from drying. The elution could be carried out under vacuum or using a syringe, but we recommend to do it by gravity for optimal peptide recovery.
9. Watch the voltage at the method start; it should not increase over 1,500 V (the electrodes might not be correctly contacting) or below 100 V (lower values could indicate too high salt concentration).
10. For instrument calibration a static off-line nanospray source is recommended to minimize the possibility of cross-contamination. The sample is loaded onto a metal-coated pulled glass capillary tip with an internal diameter (ID) ranging from 1 to 4 μm. Once the spray is stable, use the following singly charged, positive ions: 195 *m/z* for caffeine; 524 *m/z* for MRFA; and 1,222, 1,522, and 1,822 *m/z* for Ultramark 1621.
11. For instrument tune, infuse the tune solution into the MS detector at 5 μL/min. The recommended scan parameters are as follows: 400–1,100 *m/z* mass range, normal scan rate, enhanced scan type, ten microscans, and maximum injection time of 200.
12. X_{corr} is a database-independent parameter that is dependent on the quality of the tandem mass spectrum and the quality of its fit to the theoretical spectrum. ΔCn is the normalized score obtained from the difference between the first- and the second-ranked sequences. This value is highly dependent on database size, search parameters, and sequence homologies (11).
13. It is advisable that peptide quantification is based exclusively on ZoomScan spectra corresponding to peptide matches with a *FDR* lower than 5 %.
14. A mathematical analysis of the kinetic analysis shows that the three chemical species in sample *B* can be expressed as a function of the *labeling efficiency f* (16): $B_2 = B{\cdot}f^2$; $B_0 = B{\cdot}(1-f)2$; $B_1 = B{\cdot}2{\cdot}f(1-f)$.
15. Isotopic peaks were modeled using a mixed Gaussian/double exponential distribution,

$$I(x) = (1-\beta)\frac{1}{\sqrt{2\pi\sigma}}\exp\left(-\frac{(x-\mu)^2}{\sigma^2}\right) + \beta\frac{1}{2\sigma}\exp\left(-\left|\frac{x-\mu}{\sigma}\right|\right),$$

where $I(x)$ is the intensity at m/z x for a peak whose area is the unity, μ is the location parameter (the mean in the Gaussian distribution), σ is the scale parameter (the standard deviation in the Gaussian distribution, which determines the half-width of the peak), and β evaluates the relative proportion of double exponential component within the mixed distribution.

16. The best-fit values of the parameters σ and β and of the labeling efficiency f were used to eliminate unreliable quantifications. ZoomScans were automatically considered unreliable and were eliminated when the best-fit parameters verified any of the following conditions: $\sigma > 0.12$, $\beta > 1.1$, and $f > 1.2$.
17. Quantitative stable isotopic labeling experiments are peptide-centric approaches where ions are directly quantified in the MS detector; since not all spectra produce equally precise quantifications, each spectrum has a different variance, and hence quantitative data cannot be treated as a whole by using a unique normal distribution.

References

1. Assma G et al (1988) The Prospective Cardiovascular Munster (PROCAM) study: prevalence of hyperlipidaemia in persons with hypertension and/or diabetes mellitus and the relationship to coronary heart disease. Am Heart J 116:1713e24
2. Tall A (2008) Cholesterol efflux pathways and other potential mechanisms involved in the athero-protective effect of high density lipoproteins. J Intern Med 263:256e73
3. Vaisar T et al (2007) Shotgun proteomics implicates protease inhibition and complement activation in the anti-inflammatory properties of HDL. J Clin Invest 117:746–756
4. Navab M et al (2001) HDL and the inflammatory response induced by LDL-derived oxidized phospholipids. Arterioscler Thromb Vasc Biol 21:481–488
5. Bonzon-Kulichenko E et al (2011) A robust method for quantitative high-throughput analysis of proteomes by 18O labeling. Mol Cell Proteomics 10:M110.003335, Epub 2010 Aug 31
6. Jorge I et al (2007) High-sensitivity analysis of specific peptides in complex samples by selected MS/MS ion monitoring and linear ion trap mass spectrometry: application to biological studies. J Mass Spectrom 42:1391–1403
7. Ortega-Perez I et al (2005) c-Jun N-terminal kinase (JNK) positively regulates NFATc2 transactivation through phosphorylation within the N-terminal regulatory domain. J Biol Chem 280:20867–20878
8. Martinez-Ruiz A et al (2005) S-nitrosylation of Hsp90 promotes the inhibition of its ATPase and endothelial nitric oxide synthase regulatory activities. Proc Natl Acad Sci USA 102:8525–8530
9. Lopez-Ferrer D et al (2006) Quantitative proteomics using 16O/18O labeling and linear ion trap mass spectrometry. Proteomics 6(Suppl 1):S4–S11
10. Yates JR 3rd, Eng JK, McCormack AL (1995) Mining genomes: correlating tandem mass spectra of modified and unmodified peptides to sequences in nucleotide databases. Anal Chem 67:3202–3210
11. Lopez-Ferrer D et al (2004) Statistical model for large-scale peptide identification in databases from tandem mass spectra using SEQUEST. Anal Chem 76:6853–6860
12. Storey JD, Tibshirani R (2003) Statistical significance for genomewide studies. Proc Natl Acad Sci USA 100:9440–9445
13. Martinez-Bartolome S et al (2008) Properties of average score distributions of SEQUEST: the probability ratio method. Mol Cell Proteomics 7:1135–1145
14. Navarro P, Vazquez J (2009) A refined method to calculate false discovery rates for peptide identification using decoy databases. J Proteome Res 8:1792–1796
15. Elias JE, Gygi SP (2007) Target-decoy search strategy for increased confidence in large-scale protein identifications by mass spectrometry. Nat Methods 4:207–214
16. Ramos-Fernandez A, Lopez-Ferrer D, Vazquez J (2007) Improved method for differential expression proteomics using trypsin-catalysed 18O labeling with a correction for labeling efficiency. Mol Cell Proteomics 6:1274–1286
17. Yao X, Afonso C, Fenselau C (2003) Dissection of proteolytic 18O labeling: endoprotease-catalysed 16O-to-18O exchange of truncated peptide substrates. J Proteome Res 2:147–152
18. Jorge I et al (2009) Statistical model to analyse quantitative proteomics data obtained by 18O/16O labeling and linear ion trap mass spectrometry: application to the study of vascular endothelial growth factor-induced angiogenesis in endothelial cells. Mol Cell Proteomics 8:1130–1149

Chapter 12

Unraveling Biomarkers of Abdominal Aortic Aneurisms by iTRAQ Analysis of Depleted Plasma

Enrique Calvo, Roxana Martínez-Pinna, Priscila Ramos-Mozo, C. Pastor-Vargas, Emilio Camafeita, Jesús Egido, José Luis Martin-Ventura, and Juan Antonio Lopez

Abstract

Abdominal aortic aneurysm (AAA) is a significant health problem in Western countries. The diameter of AAA is a surrogate marker of its growth rate that reflects the magnitude of the degenerative process in the vascular wall, although most AAAs show discontinuous growth patterns and alternate periods of stability and nongrowth with periods of acute expansion and occasionally ruptures. Thus, the identification of biomarkers of AAA in plasma could aid in the diagnosis, prognosis, and therapy of AAA progression. However, owing to the complex composition of plasma, depletion methods must be applied before the analytical approaches for detecting low-abundant plasma protein components. In the present work, we describe a proteomic study on MARS 14-depleted plasma based on mass spectrometry (MS) which combines peptide labelling (isobaric tagging for relative and absolute quantification, iTRAQ) and liquid chromatography–tandem mass spectrometry (LC–MS/MS). This quantitative approach revealed altered levels of several proteins related to the complement system in AAA patients.

Key words Abdominal aortic aneurysm, iTRAQ, Mass spectrometry, Proteomics, Depleted plasma, Biomarkers

1 Introduction

High-throughput "omics" in systems biology approaches will ameliorate both prediction and treatment of patients on the way to the era of individualized medicine (1–3). Human plasma is the most clinically valuable specimen because it is routinely sampled with minimal invasiveness and constitutes a rich source of proteins since it is in contact with all tissues in the body. Plasma proteomics has gained much attention in the last years, despite the high complexity of the human plasma-derived proteome. Plasma samples display an extremely wide concentration range of protein components in which a set of high-abundance proteins often hinder the detection of

Fernando Vivanco (ed.), *Vascular Proteomics: Methods and Protocols*, Methods in Molecular Biology, vol. 1000, DOI 10.1007/978-1-62703-405-0_12,

low-abundance proteins with much higher biological relevance. Therefore, the initial step in most analytical approaches applied to serum or plasma is to deplete as many high-abundance proteins as possible (4). These methods facilitate plasma analysis mainly by reducing the dynamic range of protein concentration or by enriching for potentially interesting subgroups of proteins (5). Prefractionation methods can be carried out either at the protein or the peptide level, or at both. At the protein level, immunoaffinity capture using antibodies against the most abundant proteins, such as albumin or globulins, is the most popular depletion method (6). One of these systems is the Multiple Affinity Removal System (MARS) from Agilent Technologies, which offers reproducibility and sensitivity levels adequate for quantitative candidate biomarker discovery. In the present study, we described the analysis of depleted plasma from AAA patients in a nontargeted, quantitative analytical approach based on iTRAQ labelling, coupled to nLC–MS/MS (7). The depletion of the 14 most abundant proteins present in human plasma using MARS14, combined with the aforementioned MS-based strategy, enabled the reliable identification of a large number of proteins in patient and control plasma samples, while only a few peptides from proteins previously depleted by affinity chromatography were found. Quantitative data evidenced the overexpression of several proteins related to the complement system in plasma samples from patients in early disease stage, which suggests a potential role of complement proteins in the evolution of AAA.

2 Materials

All reagents are purchased from Sigma-Aldrich, unless noted otherwise, and are prepared with ultrapure water (Millipore). For HPLC and MS, all the solvents used are MS grade.

2.1 Plasma Depletion and Quantitation

1. Spin-X Centrifuge 0.45 μm filters.
2. MARS 14 High-Abundant Protein Depletion Column (or similar depletion kit): HPLC MARS Buffer A (Buffer A is a salt-containing neutral buffer, pH 7.4, used for loading, washing, and re-equilibrating the column). HPLC MARS Buffer B (Buffer B is a low-pH urea buffer used for eluting the bound high-abundant proteins from the column), all from Agilent Technologies.
3. RC-DC Protein Assay Kit (Bio-Rad).

2.2 Trypsin Digestion and iTRAQ Labelling

1. iTRAQ Reagent Kit, 4-plex (Applied Biosystems), which contains dissolution buffer, 500 mM triethylammonium bicarbonate (TEAB, pH 8.5), Denaturant solution, which contains 2 % (w/v) SDS, Reducing agent, 50 mM

Tris-(2-carboxyethyl)phosphine (TCEP), Cysteine-blocking reagent, 200 mM methyl methanethiosulfonate (MMTS) in isopropanol. Single iTRAQ reagents (114–117).

2. Trypsin (Sequence Grade Modified Trypsin, Promega).

2.3 LC–MS/MS Analysis

1. Pre-column for peptide online desalting: C-18 reversed-phase (RP) micro-column (300 mm ID × 5 mm PepMap™) (LC Packings).
2. Analytical RP column: self-packed 15-cm-long C18, 100 μm ID column (Mediterranean sea, Teknokroma).
3. Picotip emitter, 15 mm tip (New Objective).
4. HPLC Buffer A: 0.5 % (v/v) acetic acid in water.
5. HPLC Buffer B: 0.5 % (v/v) acetic acid in 90 % (v/v) acetonitrile.
6. Software for analysis and quantitation: Proteome Discoverer (Thermo Fisher Scientific), Mascot (Matrix Science), and Scaffold (Proteome Software).

2.4 Specialized Equipment

1. Agilent HPLC or similar.
2. Lyophilizer.
3. Ultimate 3000 nano-HPLC System (or equivalent) (Dionex Corporation).
4. LTQ-Orbitrap XL hybrid Linear Ion Trap Mass Spectrometer (or equivalent) (Thermo Fisher Scientific).

3 Methods

A schedule of the general procedures and multitagging experimental design is shown on Fig. 1.

3.1 Plasma

1. Collect, into EDTA tubes, three blood samples of each two groups of AAA patients: small (aaa, diameter = 3–5 cm) and large (AAA, diameter > 5 cm), and three healthy age- and sex-matched controls (*see* **Note 1**).
2. Store blood samples vertically at 4 °C for 30 min, and centrifuge blood at 3,000 × *g* for 20 min at 4 °C.
3. Store plasma samples aliquots at −80 °C for further analysis (*see* **Note 2**).

3.2 High-Abundant Protein Depletion from Plasma

1. Deplete plasma samples at room temperature with an Agilent 1100 HPLC system using a 10 × 100 mm MARS14 column according to the manufacturer's instructions (*see* **Note 3**).

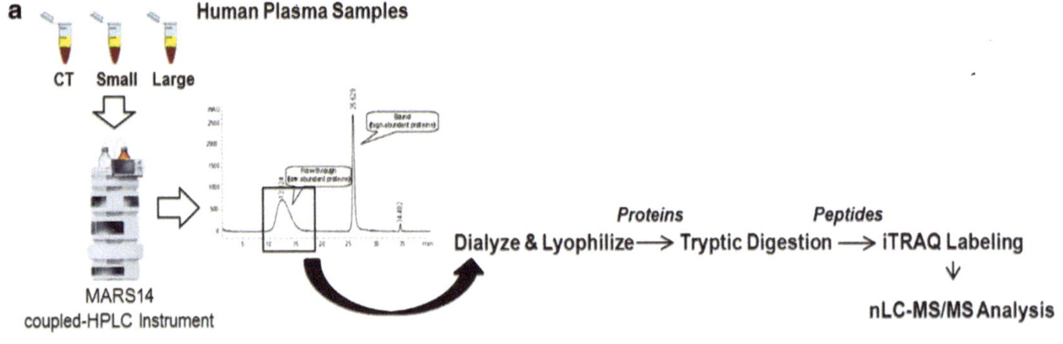

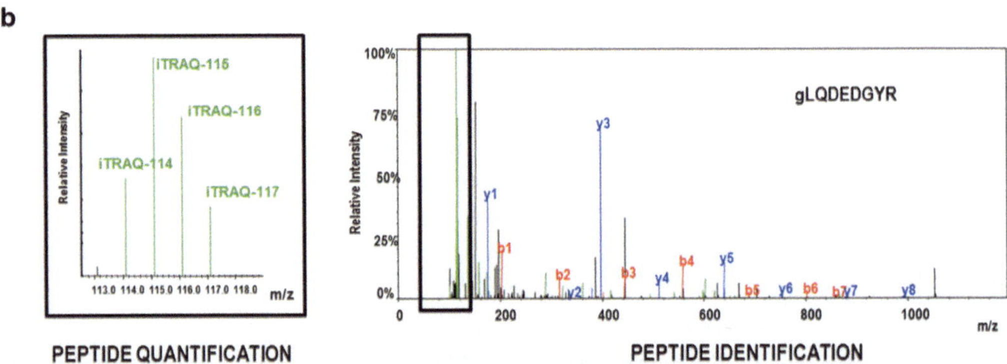

Fig. 1 Schematic of multitagging experimental strategy used in the analysis. (**a**) Plasma samples from six patients (small, three patients with small aaa; large, three patients with large AAA) and three control subjects (CT) were depleted using a MARS 14 column. Flow-through fractions containing low-abundant proteins were trypsin digested; the resulting peptides were labelled using 114–117 iTRAQ tags and, finally, analyzed by nLC–MS/MS in a LTQ Orbitrap XL mass spectrometer. (**b**) A fragmentation spectrum of a representative peptide from complement protein 4B with sequence assignment is shown; a detailed inspection of the iTRAQ reporter ion profiles in the low *m*/*z* region for protein quantification is also shown

2. Attach the MARS column and equilibrate for 4 min, with 4 mL of HPLC MARS Buffer A at a flow rate of 1 mL/min (*see* **Note 4**).
3. Dilute 200 μL of human plasma with 600 μL of HPLC MARS Buffer A (*see* **Note 5**).
4. Centrifuge diluted plasma through a Spin-X Centrifuge 0.45 μm filter, 1 min at 16,000 × *g* to remove particulates.
5. Inject the diluted sample using a flow rate of 0.5–1 mL/min (*see* **Note 6**).
6. Collect the flow-through fraction in the 10.5–14.5 min retention time range. This corresponds to the low-abundant protein fraction.
7. Dialyse the flow-through fractions (four fractions per sample) with ammonium bicarbonate to exchange buffer and then lyophilize them. Store at −80 °C for later analysis (*see* **Note 7**).

8. Elute bound proteins by adding HPLC MARS Buffer B at a flow rate of 3 mL/min during 7.5 min.
9. Regenerate column using HPLC MARS Buffer A at a flow rate of 3 mL/min for 10 min and store column with Buffer A at 2–8 °C.
10. Resuspend lyophilized fractions in TEAB Buffer.
11. Determine protein concentration by RC-DC protein assay, using bovine albumin diluted in TEAB Buffer to prepare the standard curve (*see* **Note 8**).

3.3 Trypsin Digestion

1. Pipette 50 μg of proteins from the four samples into four separate tubes (*see* **Note 9**). Make up to 19 μL with TEAB.
2. Add 1 μL of the Denaturant solution (final 0.1 % SDS) to each tube. Mix by vortexing and spin down briefly to collect the sample to the bottom of the tube (*see* **Note 10**).
3. Add 2 μL of Reducing reagent and vortex. Incubate at 60 °C for 1 h.
4. Add 1 μL of the Cysteine-blocking reagent. Incubate at room temperature for 30 min in the dark.
5. Reconstitute one vial of trypsin (20 μg) with 45 μL of MS-grade water. Mix briefly avoiding formation of foam.
6. Add 10 μL of the trypsin solution (1:10 enzyme-to-protein ratio) and vortex. Incubate overnight at 37 °C (*see* **Note 11**).
7. Transfer the resulting digestion solutions to small RT-PCR tubes and keep at 4 °C for iTRAQ labelling, or store at –80 °C.

3.4 iTRAQ Labelling

1. Take the iTRAQ reagents to room temperature. Add 70 μL of ethanol into each reagent vial, cap the vial, and vortex vigorously, and then centrifuge briefly to settle the iTRAQ reagents at the vial bottom.
2. Transfer the contents of each vial to the appropriate sample tube and mix well. For example, peptides derived from two control samples are labelled with iTRAQ Reagents 114 and 115, whereas peptides obtained from AAA and aaa patient samples are labelled with iTRAQ Reagents 116 and 117, respectively (*see* **Note 12**).
3. Transfer the entire content of one iTRAQ reagent vial to each of the four sample tubes and mix (*see* **Note 13**).
4. Incubate the reaction vials at room temperature for 2 h.
5. Analyze a small fraction of each reaction to check labelling efficiency (*see* **Note 14**).
6. If the labelling is complete, stop reaction with 3 μL glacial acetic acid and combine labelled samples. Concentrate samples in the Speedvac to remove solvent (*see* **Note 15**).

3.5 Reversed-Phase Liquid Chromatography

Use a reversed-phase cartridge for desalting the peptides and removing excess iTRAQ reagent. Then analyze the peptides by nano-electrospray ionization after separation by nano-HPLC.

1. Redissolve peptides in HPLC Buffer A and remove particulate material by centrifugation (10 min at 15,000×*g*).
2. Load peptides onto a homemade reversed-phase C18 cartridge for peptide desalting (*see* **Note 16**).
3. Elute peptides in 50 % HPLC Buffer, dry down samples in a vacuum centrifuge, and resuspend peptides in 20 μL of HPLC Buffer A just before MS analysis.
4. Equilibrate the analytical HPLC column with 3 % HPLC Buffer B for at least 15 min prior to sample injection.
5. Resolve the bound peptides on a C-18 reversed-phase (RP) nano-column to an emitter nanospray needle for real-time ionization and peptide fragmentation, at a flow rate of 300 nL/min using the gradient 0–50 % B in 90 min and 50–90 % B in 1 min (*see* **Note 17**).

3.6 Mass Spectrometry

Tandem mass spectrometry analysis is performed using an LTQ Orbitrap XL hybrid mass spectrometer. The MS is operated in positive ion mode with a capillary temperature of 200 °C with no sheath gas.

1. Acquire full scan MS in profile mode from 390 to 1,200 *m*/*z* at 30,000 resolution using an internal lock mass. Maximal accumulation time for MS1 is set to 390 ms to a target fill of 1×10^{6}.
2. A minimum signal of 1,000 counts is required to trigger data-dependent acquisition of MS/MS spectra. These are acquired in a data-dependent mode consisting of selection of the three most abundant ions on each cycle. For iTRAQ analysis, alternate between collision-induced disassociation (CID) and higher energy dissociation (HCD) modes for each selected precursor ion, so only three precursor ions are analyzed per cycle. For both fragmentation modes, accumulate to 5×10^{4} ions (*see* **Note 18**).
3. Dynamic exclusion parameters allow two repeat hits before the precursor m/z is added to an exclusion list for 180 s.

3.7 Database Searching and iTRAQ Labelling Analysis

1. Extract mass spectra by using Proteome Discoverer v1.2 software. For peptide identification, search the MS/MS spectra against a human protein database subset of the IPI database (human_ref.fasta; 2003, April; 39414 entries; *see* **Note 19**) using Sequest (Thermo Fisher version 1.0.43.2) and Mascot (local version 2.1.) engines. Use the following search parameters:

trypsin digestion with up to two missed cleavages; mass tolerance of 15 ppm and 0.8 Da for precursor and fragment search, respectively; MMTS-labelled cysteines and iTRAQ-labelled N-termini as fixed modifications; and phosphorylation of serine, threonine, and tyrosine and iTRAQ labelling of threonine and lysine residues as variable modifications. Corrections were applied for impurity of iTRAQ reagents according to the manufacturer's instructions (*see* **Note 20**).

2. Use Scaffold v.3.00.02 for validation and quantitation (*see* **Note 21**). Consider peptides identified with a probability greater than 80 % and accept proteins identified at greater than 80 % probability with at least one peptide identified, as specified by the Peptide and Protein Prophet (8) algorithms, respectively.
3. For protein quantification, decrease the minimum peptide and protein identification probabilities to 0.0 and 20 %, respectively, to include the low quality HCD spectra containing iTRAQ reporter quantitative data.
4. Quantify proteins containing at least two matched peptides to reduce the probability of a false-positive identification.
5. Highlight differentially expressed proteins using two criteria: (i) p values ≤ 0.05 and (ii) fold change value higher than 2 (*see* **Note 22**).

4 Notes

1. Written consent from all the individuals must be obtained, following the guidelines of the Ethics Committee of your institution. All subjects are in a fasting state; blood samples are taken in the morning (between 8 and 11 h) in an EDTA vacutainer.
2. Protein concentration can be measured at this step to determine the volume of plasma proteins to be loaded into the depletion column.
3. Depletion of high-abundant proteins can be performed using a variety of commercially available depletion technologies. The method described here is based on the MARS 14 column (4.6 × 100 mm), designed to remove 14 abundant proteins based on affinity interactions (albumin, IgG, antitrypsin, IgA, transferrin, haptoglobin, fibrinogen, alpha2-macroglobulin, alpha1-acid glycoprotein, IgM, apolipoprotein AI, apolipoprotein AII, complement C3, and transthyretin). While the high-abundant proteins are specifically retained into the column, the low-abundant proteins are collected to the flow-through fractions for further analysis.

4. The LC system must be purged before attaching the column for 10 min using HPLC MARS Buffer A at a flow rate of 1 mL/min, and set up HPLC MARS Buffer A and B as the only mobile phases.
5. It is recommended to dilute samples four times with HPLC MARS Buffer A before loading to facilitate protein binding.
6. Let samples interact with the stationary phase for complete binding of high-abundant proteins using a low flow rate.
7. It is important to carry out a thorough desalting, as iTRAQ labelling could be hampered by the presence of unwanted components like primary amines from Tris–HCl buffer.
8. If the lysis buffer contains detergents or reducing agents, RC-DC protein assay must be used for accurate protein concentration estimation.
9. It is essential to use comparable protein amounts in the four samples to facilitate later data analysis and to ensure adequate iTRAQ labelling; excessive protein amounts will result in incomplete labelling and/or bias in the quantification.
10. Mix well throughout the whole protocol: add carefully the components, vortex, and briefly centrifuge to spin down the sample to the bottom of the tubes.
11. To avoid partial digestion, overnight incubation is recommended. If possible, check protein digestion efficiency by analyzing a minute amount of the digested sample: desalt it using a ZipTip (Millipore), mix the eluted peptides with MALDI matrix solution, and spot them onto a MALDI plate (Bruker Daltonik). Check the MS spectra for comparable peptide ion signals between samples.
12. Based on iTRAQ instructions from ABI, the concentration of ethanol and iTRAQ reagents in the iTRAQ labelling reactions should be larger than 60 % (v/v) to maximize labelling efficiency.
13. Based on iTRAQ instructions from ABI, each iTRAQ vial can be used to label up to 100 μg of protein.
14. Because partial labelling is feasible, it is advisable to test the labelling efficiency in a small peptide sample before mixing using MALDI MS as in Note 11.
15. To completely remove TEAB, reconstitute iTRAQ-labelled samples in 100 μL of MS-grade water and dry the sample in a vacuum concentrator. Repeat this step twice to ensure complete evaporation of TEAB.
16. Despite that a RP-C18 cartridge was employed for peptide desalting, the use of strong cation exchange (SCX) cartridges or columns for peptide desalting is widespread. Furthermore,

depending of sample complexity and the amount of labelled protein, additional fractionation steps may be necessary to effectively reduce peptide complexity (e.g., by SCX chromatography).

17. The gradient used depends on the complexity of the peptide mixture and the available HLPC.
18. Different MS instruments have different optimal MS and MS/MS methods for the analysis of iTRAQ-labelled peptides (see refs. 9–11).
19. It is recommended to use the latest version of the protein database (i.e., Swissprot, IPI, or NCBI protein database) to ensure comprehensive peptide identification. In this study the IPI database was chosen owing to its high number of protein sequence entries, but this database has been recently discontinued.
20. If too many variable modifications are considered, confidence interval values for spectra matching can decrease due to the increased random matches. It is preferable to include only those modifications relevant to the experiment.
21. There are many other programs for the analysis of iTRAQ data (e.g., ref. 12).
22. Student's *t* test is utilized to account for the relative protein expression changes between groups for each protein. It is advisable to employ several biological and control sample replicates for increasing the statistical confidence of analysis. It is very important to verify the iTRAQ results using biological complementary techniques, as Western blotting, ELISA, or immunohistochemistry.

Acknowledgments

The work has been supported by the Spanish MICIN (SAF2010/21852), Redes RECAVA (RD06/0014/0035), EUS2008-03565, and Fundacion Pro-CNIC. The Centro Nacional de Investigaciones Cardiovasculares is supported by the Spanish Ministry of Science and Innovation (MICIN) and the Pro-CNIC Foundation.

References

1. Sakalihasan N, Limet R, Defawe OD (2005) Abdominal aortic aneurysm. Lancet 365:1577–1589
2. Vega de Céniga M, Gómez R, Estallo L, de la Fuente N, Viviens B, Barba A (2008) Analysis of expansion patterns in 4-4.9 cm abdominal aortic aneurysms. Ann Vasc Surg 22:37–44
3. Martinez-Pinna R, Barbas C, Blanco-Colio LM, Tunon J, Ramos-Mozo P, Lopez JA, Meilhac O, Michel JB, Egido J, Martin-Ventura JL (2010) Proteomic and metabolomic profiles in atherothrombotic vascular disease. Curr Atheroscler Rep 12: 202–208

4. Pernemalm M, Lewensohn R, Lehtiö J (2009) Affinity prefractionation for MS-based plasma proteomics. Proteomics 9:1420–1427
5. Polaskova V, Kapur A, Khan A, Molloy MP, Baker MS (2010) High-abundance protein depletion: comparison of methods for human plasma biomarker discovery. Electrophoresis 31:471–482
6. Gong Y, Li X, Yang B, Ying W, Li D, Zhang Y, Dai S, Cai Y, Wang J, He F, Qian X (2006) Different immunoaffinity fractionation strategies to characterize the human plasma proteome. J Proteome Res 5: 1379–1387
7. Treumann A, Thiede B (2010) Isobaric protein and peptide quantification: perspectives and issues. Expert Rev Proteomics 7: 647–653
8. Nesvizhskii AI, Keller A, Kolker E, Aebersold R (2003) A statistical model for identifying proteins by tandem mass spectrometry. Anal Chem 75:4646–4658
9. Mischerikow N, van Nierop P, Li KW, Bernstein HG, Smith AB, Heck AJ, Altelaar AF (2010) Gaining efficiency by parallel quantification and identification of iTRAQ-labelled peptides using HCD and decision tree guided CID/ETD on an LTQ Orbitrap. Analyst 135:2643–2652
10. Bantscheff M, Boesche M, Eberhard D, Matthieson T, Sweetman G, Kuster B (2008) Robust and sensitive iTRAQ quantification on an LTQ Orbitrap mass spectrometer. Mol Cell Proteomics 7:1702–1713
11. Dayon L, Pasquarello C, Hoogland C, Sanchez JC, Scherl A (2010) Combining low- and high-energy tandem mass spectra for optimized peptide quantification with isobaric tags. J Proteomics 73:769–777
12. Li Z, Adams RM, Chourey K, Hurst GB, Hettich RL, Pan C (2012) Systematic comparison of label-free, metabolic labeling, and isobaric chemical labeling for quantitative proteomics on LTQ Orbitrap Velos. J Proteome Res 11(3):1582–1590

Chapter 13

Absolute Quantitation of Proteins in Human Blood by Multiplexed Multiple Reaction Monitoring Mass Spectrometry

Andrew J. Percy, Andrew G. Chambers, Carol E. Parker, and Christoph H. Borchers

Abstract

Multiple reaction monitoring (MRM)-mass spectrometry (MS) with stable isotope-labeled standards (SIS) has proven adept in rapidly, precisely, and accurately quantifying proteins in complex biological samples. The impetus behind the early use of multiplexed MRM in proteomics was to expedite the verification and validation stages of the protein biomarker pipeline for clinical utility, which involves the analysis of hundreds or even thousands of samples. Moreover, once a multiplexed assay has been developed, however, it can be turned around and used for biomarker discovery, as has been demonstrated for cancer biomarkers by our laboratory and by others. Overall, these MRM-based methods compare favorably with antibody-based techniques, such as ELISAs or protein arrays, in that MRM-based methods are less expensive and can be developed more rapidly.

There are two MRM-based platforms that are currently being developed: a standard-flow and a nano-flow LC/ESI-MRM-MS (liquid chromatography-electrospray ionization) platform. In this book chapter, we describe a recent study in which we evaluated these two platforms, both interfaced to the same mass spectrometer. This study demonstrated the enhanced performance metrics (in terms of sensitivity, dynamic range, and robustness) of the standard-flow ultra-high performance liquid chromatography (UHPLC) system compared to the nano-flow HPLC-Chip for the absolute quantitation of 48 plasma proteins. Using the standard-flow platform, we also developed two high-throughput assays for the analysis of a panel of 67 cardiovascular disease (CVD) biomarkers in non-depleted and non-enriched human plasma and a panel of 25 putative biomarkers in dried human blood spots (DBS). Since the nanoLC/MRM-MS platform has advantages under sample-limited conditions and for the analysis of certain specific peptides, the protocols for both systems are described here.

Key words Plasma, Dried blood spot (DBS), Protein, Stable isotope-labeled standards (SIS), Multiple reaction monitoring (MRM), Mass spectrometry (MS), Quantitative proteomics

1 Introduction

With the growing prevalence of cardiovascular disease (CVD) in today's population, a sizeable amount of research effort has been devoted to early detection for improved disease management.

Fernando Vivanco (ed.), *Vascular Proteomics: Methods and Protocols*, Methods in Molecular Biology, vol. 1000, DOI 10.1007/978-1-62703-405-0_13,

This has enabled the discovery of hundreds of potential protein indicators (aptly termed biomarkers) in human bodily fluids, such as blood plasma (1–3). Blood plasma is a popular biofluid for screening as it is easy to obtain, inexpensive to sample, and contains a repository of proteins that originate from circulating blood cells and surrounding tissues. Of the 177 proposed biomarkers found in human plasma (1), only a small number have been verified and even fewer validated (2), which requires testing the markers on extremely large cohorts of patient samples in order to certify them as true clinical biomarkers. Translation of putative protein biomarkers to clinical application therefore requires a high-throughput and robust platform in order to screen thousands of patient samples, ideally, in a multiplexed manner to reduce processing time and cost. The multiple reaction monitoring/mass spectrometry (MRM-MS) approach with stable isotope-labeled standards (SIS), which has long been utilized for small molecule quantitation (4, 5), satisfies the throughput and multiplexing requirements and is emerging as a viable technique for determining plasma concentrations of CVD-related protein biomarkers in human biofluids (6–8). This approach is based on the measurement of proteotypic peptides which function as molecular surrogates for the proteins of interest and uses SIS peptides as internal standards for normalization and peak verification. For highly specific and sensitive quantitative measurements, the MRM mode of MS requires a priori knowledge of the analytes and judicious selection of both the SIS peptides and the MRM precursor/product ion pairs.

At present, standard-flow and nano-flow LC/MRM-MS methods can quantitate low ng/mL levels in blood plasma relatively routinely, if partnered with up-front immunoaffinity depletion (9–11) or enrichment (12, 13). Such prefractionation, however, compromises the intra-sample reproducibility and eliminates the diagnostically useful high-abundance proteins. Furthermore, it remains unclear as to what LC-MS platform is analytically superior due to the dissimilarity in the variables and conditions utilized between methods. To this end, we compared the performance metrics of a standard-flow UHPLC system to a nano-flow HPLC-Chip system, both interfaced to the same state-of-the-art triple quadrupole (QqQ) mass spectrometer (Agilent's 6490), for the directed quantitation of 48 CVD-associated proteins in non-depleted and non-enriched human plasma. This instrument was selected for this comparison study as it has demonstrated enhanced detection sensitivity and contamination removal due to a unique front-end design that is centered on iFunnel technology.

The comparison study revealed that the standard-flow LC/MRM-MS platform had reduced retention time variability and superior sensitivity (14). This standard-flow platform was then applied to the development of two high-throughput multiplexed MRM-based methods for quantitating 67 and 25 high-to-moderate abundance candidate CVD-related protein biomarkers in non-depleted and non-enriched human plasma (15) and human dried blood spots (DBS) (16), respectively. DBS sampling enables small volumes of blood to be readily collected and often improves analyte stability at room temperature storage. These advantages have led to the development of DBS sampling methods coupled with MRM-MS for analysis of small molecules for newborn screening of metabolism disorders and for preclinical toxicology studies (17, 18). Most current methods for protein quantitation in DBS are optimized for one or two targeted proteins, but a general method for processing and analyzing proteins from DBS samples would support the development of highly multiplexed assays.

This chapter describes the necessary materials and quantitative proteomic methods we use to perform standard-flow and nanoflow LC/MRM-MS analyses of human plasma and human DBS. We conclude with a discussion on the significance of these applications to the biomedical research field. Overall, the methods are robust, sensitive, and provide the throughput and multiplexing capabilities required to verify and validate high-to-moderate abundance putative CVD-linked biomarkers in human blood.

2 Materials

Unless otherwise stated, all chemicals and reagents are purchased from Sigma-Aldrich (St. Louis, MO, USA) and are the highest grade available. All solutions are prepared from LC-MS grade solvents (Sigma-Aldrich).

2.1 Biological Samples

1. Human plasma (Bioreclamation; Westbury, NY, USA; part no. HMPLEDTA2).
2. Human whole blood (Bioreclamation; part no. HMWBEDTA2).

2.2 Dried Blood Spot Processing

1. DBS collection card (903 Protein Saver Card; Whatman; NJ, USA).
2. Hermetically sealed bag (Ziploc).
3. Sodium calcium aluminosilicate hydrate desiccant (VWR International; West Chester, PA, USA; part no. 61161-321).
4. Humidity indicator card (VWR International; part no. 4HIC100).
5. Eppendorf Thermomixer R with block for 1.5 mL tubes (Brinkmann Instruments; Westbury, NY, USA; part no. 022670107 and 022670522).

2.3 Sample Preparation

1. Ammonium bicarbonate.
2. Sodium deoxycholate.

3. Tris(2-carboxyethyl)phosphine (Thermo Scientific; Rockford, Il, USA).
4. Iodoacetamide.
5. Dithiothreitol.
6. Trypsin, modified porcine sequencing grade (Promega; Madison, WI, USA).
7. SIS peptides.
8. Formic acid.
9. Methanol.
10. 10 mg Oasis HLB cartridge (part no. 186000383; Waters; Milford, MA, USA).

2.4 LC/MRM-MS Systems

1. 1290 Infinity UHPLC system for standard-flow experiments (Agilent Technologies; Santa Clara, CA, USA).
2. 1200 series nanoLC system for nano-flow experiments (Agilent Technologies).
3. Reversed-phase UHPLC column (part no. 959759-902; 150×2.1 mm i.d., Zorbax RRHD Eclipse Plus C_{18}, 1.8 μm particles; Agilent Technologies).
4. Reversed-phase HPLC-Chip column (ProtID-Chip-150, part no. G4240-62006; 150 mm×75 μm, i.e., Zorbax 300SB-C_{18}, 5 μm particles; Agilent Technologies).
5. Dedicated infusion and flow injection chip (FIA-Chip (II), part no. G4240-61015; Agilent Technologies).
6. LC gradient mobile phase A: 0.1 %v/v formic acid.
7. LC gradient mobile phase B: 90 %v/v acetonitrile with 0.1 %v/v formic acid.
8. Standard-flow ESI (Jet Stream) source (Agilent Technologies) or nano-flow ESI interface (Agilent Technologies).
9. Chip Cube (part no. G4240A; Agilent Technologies).
10. QqQ 6490 mass spectrometer (Agilent Technologies).
11. Harvard PicoPlus 11 syringe pump (Harvard Apparatus; Holliston, MA, USA)
12. Hamilton gastight syringe (1 mL, 22 s gauge, 2 in. length, point style 3).

2.5 Data Analysis Software

1. Mass Hunter Qualitative Analysis (Agilent Technologies).
2. Mass Hunter Quantitative Analysis (Agilent Technologies).
3. Mass Hunter Optimizer for Peptides (Agilent Technologies).
4. Molecular Weight Calculator.

3 Methods

The described methods were developed to quantitate panels of moderate-to-high abundance proteins in human plasma and human DBS using a standard-flow or nano-flow LC/MRM-MS platform. We used biofluids from Bioreclamation that represent pooled samples of whole blood donations collected from a group of healthy, race- and gender-matched donors between the ages of 18 and 50. The plasma samples were stored at −20 °C until use, in order to reduce degradation, while the whole blood samples are stored at 4 °C to prevent lysis of the red blood cells. Nonetheless, whole blood is perishable and should be used within the first few days of collection, if possible.

3.1 Dried Blood Spot Processing

1. Spot 15 μL of whole blood on collection cards with a calibrated pipette (*see* **Note 1**).
2. Allow the cards to dry for 3 h at room temperature.
3. Store the cards in a hermetically sealed bag with desiccant and a humidity indicator card.
4. When required for analysis, excise the entire DBS from the collection card using scissors or a large hole punch (*see* **Note 2**), and proceed to **step 2** of Subheading 3.2.

3.2 Sample Preparation

The procedures that we used for preparation of the plasma and the DBS for digestion, acidification, and extraction are similar to one another and are also similar to those previously described in the chapter by Kuzyk et al., in an earlier volume in this series (19). Briefly, the following sequential steps are conducted:

1. In a 1.5 mL polypropylene tube, whole plasma is thawed and briefly vortexed before diluting an aliquot 1:10 with 25 mM ammonium bicarbonate. Store the original and dilute sample on ice.
2. In a separate Eppendorf tube, add an aliquot of diluted plasma or an excised DBS card to sodium deoxycholate (initial 10 %w/v in 25 mM ammonium bicarbonate) to begin unfolding the proteins' tertiary structure (final concentration of sodium deoxycholate: 1 %; *see* **Note 3**). The DBS card may need to be folded to fit into the 1.5 mL tube.
3. Reduce the disulfide bonds with 50 mM Tris(2-carboxyethyl) phosphine (prepared in 25 mM ammonium bicarbonate) to give a final concentration of 5 mM.
4. Incubate the plasma samples at 60 °C for 30 min and the DBS samples at 60 °C while vortexing on the Thermomixer at 1,000 rpm for 1 h.

5. Alkylate the free sulfhydryl groups with a 100 mM iodoacetamide (prepared in 25 mM ammonium bicarbonate) addition to give a final concentration of 10 mM.
6. Incubate at 37 °C for 30 min in the dark.
7. Quench the remaining iodoacetamide by adding 100 mM dithiothreitol (prepared in 25 mM ammonium bicarbonate).
8. Incubate at 37 °C for 30 min.
9. Add modified, sequencing-grade trypsin (0.4 mg/mL in water) to the samples at a 50:1 substrate to enzyme ratio and allow digestion to proceed for 16 h at 37 °C.
10. Terminate digestion with the addition of an acidified SIS peptide mixture (in formic acid) to give a final formic acid concentration of 0.5 % v/v and a pH <3 (*see* **Note 4**). The mixture is composed of SIS peptides that are concentration-balanced, as described by Kuzyk et al. (19), to enable an approximately equal response of the SIS peptides when spiked into the endogenous (natural, NAT) tryptic digest.
11. Pellet the sodium deoxycholate precipitate by room temperature centrifugation at 12,000 × *g* for 10 min.
12. Desalt and concentrate an aliquot of the peptide supernatant by solid-phase extraction using a 10 mg Oasis HLB cartridge (*see* **Note 5**).
13. Freeze the eluted samples at −80 °C then lyophilize to dryness overnight.
14. Store the concentrated peptide lyophilizate at −80 °C or rehydrate for immediate LC/MRM-MS analysis. In rehydration, a volume of 0.1 %v/v formic acid is added to give a final concentration of ca. 1 μg/μL.

3.3 LC Separation Parameters

1. The UHPLC column is maintained at 50 °C for enhanced column efficiency and retention time stability, while the HPLC-Chip column is housed in a non-temperature-controlled unit.
2. All samples are placed in an autosampler tray within the 1290 Infinity System or the 1200 series nanoLC system for stable temperature control at 4 °C and to minimize sample degradation.
3. In the standard-flow experiments, load 10 μL of reconstituted sample onto the UHPLC column in Agilent's 1290 Infinity System with 3 % mobile phase B at a flow rate of 0.4 mL/min (*see* **Note 6**).
4. In the nano-flow experiments, wash 1 μL of loaded sample on the enrichment column of the ProtID-Chip-150 with 100 % mobile phase A at a flow rate of 2.8 μL/min before separating the peptides on the HPLC column at 300 nL/min (*see* **Notes 6** and 7).

5. Separate the samples in the standard-flow and nano-flow systems with a multistep gradient over 30 min for high-throughput analyses.
6. We use post-run times of 4 min in the standard-flow system and 6 min in the nano-flow system at 3 % mobile phase B for column re-equilibration.
7. To reduce inter-sample carryover, run one blank between biofluid samples in the nano-flow system and two blanks between the different sample concentrations in the loading capacity and the quantitation experiments in both systems.

3.4 MRM-MS Acquisition Parameters

The MRM acquisition parameters specified in this section pertain to the Agilent 6490 instrument that is controlled by Agilent's Mass Hunter Workstation software (version B.04.01).

1. Operate the MS system in the positive ion mode with a capillary voltage of 3,500 V in the standard-flow system or 1,600 V in the nano-flow.
2. Optimal MS parameters for the standard-flow measurements are 300 V nozzle voltage, 11 L/min sheath gas flow at a temperature of 250 °C, 15 L/min drying gas flow at a temperature of 150 °C, and 30 psi nebulizer gas flow. The carrier gas is ultra-high purity nitrogen.
3. Use a drying gas flow of ultra-high purity nitrogen at 11 L/min and a temperature of 150 °C in the nano-flow system.
4. For enhanced specificity in analyzing complex biological samples, use unit resolution (0.7 Da full width at half maximum, FWHM) in the first quadrupole (Q1) and the third quadrupole (Q3).
5. Use a default fragmentor voltage of 380 V and a cell accelerator potential of 5 V for all unscheduled and scheduled MRM transitions.
6. Use optimal, transition-specific collision energy (CE) voltages for collision-induced peptide dissociation (*see* Subheading 3.5.1 for optimization details).
7. Select target peptides in accordance to rules outlined in Kuzyk et al. (18), then choose target MRM Q1/Q3 ion pairs from the CE optimization (*see* Subheading 3.5.1) and the interference determination results (*see* Subheading 3.5.4).
8. Use peptide-specific retention times, as determined from the instructions listed in Subheading 3.5.2.
9. During dynamic MRM analyses, we recommend starting with transition retention time windows of 1 min (i.e., ±30 s) for standard-flow measurements and 2 min (i.e., ±1 min) for nano-flow. The latter detection window is wider to compensate for the twice as broad peaks generated from a nanoLC separation (14).

Despite this, it may be necessary to adjust the per-transition detection windows to achieve dwell times ≥10 ms, while collecting 10–15 data points over the chromatographic peak to define the peak shape.

3.5 Preliminary Experiments

Before quantitating the target proteins in the biological samples, a few preliminary experiments must be conducted. In order, these include optimizing the peptides CE, optimizing the LC gradient, optimizing the system loading capacity, and screening the target analytes for chemically interfering ions. Together, these optimizations will enable sensitive and interference-free Q1/Q3 MRM ion pairs to be monitored, which will improve the accuracy and reliability of the obtained quantitative results. The following subsections explain these optimization procedures as they pertain to our specific applications.

3.5.1 Empirical Optimization of Peptide Collision Energies

To ensure that the target product ions generate the most sensitive MRM ion pairs for each peptide monitored, the collision energies required for collision-induced peptide fragmentation must be optimized. While software and formulas are available for predicting appropriate CE values (20, 21), our laboratory has previously demonstrated that improved MRM signal intensities result from empirical tuning (22). As a consequence, we empirically optimize this MRM acquisition parameter at the outset of a given study if and only if the peptides in the target panel have not been previously optimized on the mass spectrometer employed. This is accomplished as follows:

(a) Prepare groups of ten randomly selected SIS peptide mixtures at 1 pmol/μL each in 30 % acetonitrile and 0.1 % formic acid.

(b) Use Agilent's Mass Hunter Optimizer for Peptides software to generate a list of b- and y-series product ions for both 2+ and 3+ precursor charge states spanning an m/z range from 300 to 1,400.

(c) For direct infusion, dissimilar parameters to those stated in Subheading 3.4 for standard-flow LC/MRM-MS analyses are a drying gas flow rate of 11 L/min, a nebulizer gas flow of 15 psi, a cell acceleration potential of 7 V, and enhanced resolution (0.4 Da FWHM) for Q1 and Q3.

(d) Unscheduled MRMs are acquired with dwell times of 20 ms for each transition and ramping CE values between 5 and 53 V, in 4 V increments.

(e) Load 600 μL of the diluted SIS mix into a gastight syringe for continuous and direct infusion, via a Harvard syringe pump, into a standard-flow ESI source at a flow rate of 7 μL/min or a nanoelectrospray infusion chip at 0.3 μL/min.

(f) Based on the observed abundances, generate a list of the top five signal-producing transitions and their optimized CE voltages for each SIS peptide.

3.5.2 LC Gradient Optimization

Optimizing the LC gradient is necessary to spread out the distribution of the eluted MRM ion pairs to give a reduced number of concurrent MRM transitions and therefore a reduced possibility for matrix-induced chemical interference in the MRM ion channels. This preliminary experiment is of particular importance in our applications given the complexity of human blood. Human plasma, for instance, is regarded as the most complex human proteome sample (3), as it contains in excess of 10,000 proteins with molecular weights spanning six orders of magnitude and protein concentrations spanning at least ten orders of magnitude (e.g., albumin, 41 mg/mL; and interleukin-1 beta, 1.2 pg/mL). An optimal LC gradient is therefore necessary to achieve sensitive and accurate MRM-MS measurements.

(a) Dilute the concentration-balanced SIS peptide mixture (250 fmol/μL) to 10 fmol/μL in buffer (i.e., 0.1 % formic acid).

(b) Create the MRM-MS acquisition method for the SIS and/or NAT peptides using the top signal-producing ion from the curated list of the top five MRM transitions per peptide (*see* **Notes 8** and **9**). Divide the target transitions into groups if the cycle time exceeds 1 s and the minimum dwell time is lower than 10 ms.

(c) Use the LC parameters and the initial 30 min gradient listed in **step 5** of Subheading 3.3 (10 μg NAT and/or 100 fmol SIS, standard-flow measurements; 1 μg NAT and/or 10 fmol SIS, nano-flow).

(d) Run unscheduled MRM analyses of the SIS mix prior to gradient testing the biofluid. This helps to verify the SIS and NAT retention times and also assist in scheduling the MRM transitions. For further verification of the NAT retention times, multiple transitions per peptide can be monitored; however, we usually optimize the gradient with a single transition/peptide.

(e) Start and end each LC gradient with 3 % mobile phase B and adjust the gradient accordingly through steps and ramps to disperse the MRM transition retention times (extracted ion chromatograms are visualized in Mass Hunter Qualitative Analysis; *see* **Note 10**).

(f) The final gradient that we use in the standard-flow separation of human plasma digest is as follows (time, %B): 0 min, 3 %; 0.1 min, 10 %; 3 min, 11 %; 13 min, 19 %; 13.5 min, 20 %; 13.6 min, 23 %; 16.7 min, 25 %; 19.7 min, 28.5 %; 21.7 min, 34 %; 22.5 min, 42 %; 23.5 min, 90 %; 26.9 min, 90 %; 30 min,

Table 1
Sample details for the loading capacity and LLOQ determinations from the plasma sample analyses. Tabulated are the on-column loaded amounts for the SIS and NAT peptides used in the standard-flow and nano-flow LC/MRM-MS platform comparison study. Reprinted from ref. 14, with permission

	Loading capacity				LLOQ			
	Standard flow		Nano-flow		Standard flow		Nano-flow	
Level	SIS[a]	NAT[b]	SIS[a]	NAT[b]	SIS[a]	NAT[b]	SIS[a]	NAT[b]
L1	100	0.07	10	0.03	0.1	10	0.01	1
L2	100	0.21	10	0.06	1	10	0.1	1
L3	100	0.62	10	0.12	10	10	1	1
L4	100	1.85	10	0.25	50	10	5	1
L5	100	5.56	10	0.5	100	10	10	1
L6	100	16.67	10	1	200	10	20	1
L7	100	50	10	2	500	10	50	1
L8	100	150	10	4	N/A	N/A	N/A	N/A

[a]The amount of SIS peptides in the tryptic plasma digest loaded onto the UHPLC and HPLC-Chip columns in the standard-flow and nano-flow platforms, respectively, are reported in fmol

[b]The amount of Nat peptides in the tryptic plasma digest loaded onto the UHPLC and HPLC-Chip columns in the standard-flow and nano-flow platforms, respectively, are reported in μg

3 %. The gradient we use for the standard-flow separation of the DBS digests is 0 min, 3 %; 0.1 min, 10 %; 20 min, 20 %; 25 min, 40 %; 26 min, 90 %; 29 min, 90; and 30 min, 3 %. The nano-flow gradient is 0 min, 3 %; 0.1 min, 8 %; 15 min, 35 %; 22.5 min, 42 %; 24.5 min, 90 %; 26.5 min, 90 %; and 30 min, 3 %.

3.5.3 Optimization of Sample Loading Amount

The determination of the optimal amount of protein digest to be used in the assay is an additional, yet essential, preliminary experiment. In our laboratory, we determine this amount by analyzing a series of standard samples containing a constant amount of SIS and a variable amount of NAT. The maximum amount of sample that results in a linear increase in the peak area of the NAT, on average, is considered optimal. This experiment need only be conducted once for each analytical column investigated.

(a) Prepare a series of digests containing a fixed quantity of SIS peptides and a variable amount of NAT (see Table 1).

(b) Create a dynamic MRM-MS acquisition method for the SIS and NAT peptides using the most intense signal-producing ion from the curated list of top five MRM transitions per peptide (*see* **Note 9**).

(c) Load the standard samples onto the UHPLC or HPLC-Chip column for LC/MRM-MS analyses using the optimal LC parameters and LC gradient outlined in Subheadings 3.3 and 3.5.2, respectively (*see* **Note 11**).

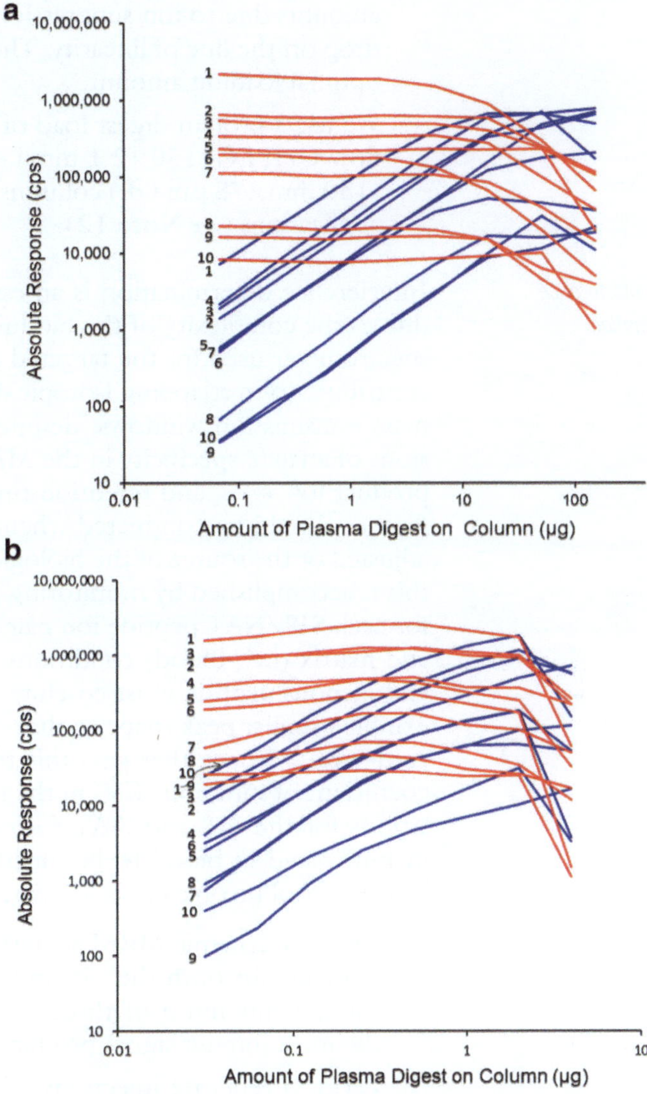

Fig. 1 Loading capacity optimization on a standard-flow UHPLC and nano-flow HPLC-Chip system. The absolute response of an identical set of representative peptides (NAT in *blue* and their corresponding SIS in *red*) is displayed as a function of the on-column amount of plasma tryptic digest. Peptide targets are for (1) α_1-antitrypsin, SVLGQLGITK; (2) ceruloplasmin, GAYPLSIEPIGVR; (3) hemopexin, NFPSPVDAAFR; (4) haptoglobin, DIAPTLTLYVGK; (5) complement C1 inactivator, FQPTLLTLPR; (6) α_1-antichymotrypsin, EIGELYLPK; (7) complement C1 inactivator, GVTSVSQIFHSPDLAIR; (8) antichymotrypsin, AVLDVFEEGTEA SAATAVK; (9) ceruloplasmin, DIFTGLIGPMK; and (10) fibrinopeptide A, SYTITGLQ PGTDYK. Reprinted from ref. 14, with permission

(d) Process the MRM data with the Mass Hunter Quantitative Analysis software using default values for peak integration.

(e) Plot the absolute response (i.e., peak area) of the NAT and SIS peptides as a function of the on-column digest amount (see Fig. 1). The SIS response will diminish at higher loading amounts due to ion suppression, while the NAT response will drop off the line of linearity. The point of deviation reflects the optimal loading amount.

(f) We use a protein digest load of 10 and 1 μg for the standard-flow UHPLC (150 × 2.1 mm i.d.) and nano-flow HPLC-Chip (150 mm × 75 μm i.d.) columns, respectively, as optimal in our applications (*see* **Note 12**).

3.5.4 Interference Determination

Interference determination is an essential preliminary experiment due to the complexity of the biofluid and the low-resolution mass spectrometer used for the targeted analysis. These two factors can contribute to overlapping isotopic distributions in the Q1 and Q3 mass transmission windows, despite the presence of three dimensions of analyte specificity in the MRM assays (precursor ion m/z, product ion m/z, and retention time). Screening for signal interference should be conducted whenever the LC gradient has been adjusted or the source of the biological sample changed. In general, this is accomplished by monitoring at least three MRM transitions for each SIS/NAT peptide ion pair under matrix-free (i.e., buffer) and matrix (i.e., blood) conditions. In comparing the signals, the endogenous peptide must co-elute with the internal standard and exhibit a similar peak shape to that of the SIS peptide to be considered "real." Those that pass this test, must also display a <20 % coefficient of variation (CV) in the relative ratio of the MRM transitions for the SIS and NAT forms of a peptide when analyzed in buffer and in blood to be qualified as being interference-free. The protocol involved is as follows:

(a) Create a dynamic MRM acquisition method containing MRM ion pairs for both the SIS and the NAT forms of the peptide using a minimum of three transitions, which correspond to the most intense signal-producing ions for each peptide.

(b) Perform replicate injections of 10 or 1 μL (standard-flow or nano-flow, respectively) of the concentration-balanced, matrix-free SIS peptide (100 fmol, standard-flow; 10 fmol, nano-flow) onto the LC/MRM-MS system and analyze for SIS.

(c) Conduct replicate LC/MRM-MS runs of SIS-spiked digest (prepared as described in Subheading 3.2) and analyze for SIS and NAT (standard-flow: 100 fmol SIS, 10 μg NAT; nano-flow: 10 fmol SIS, 1 μg NAT).

(d) Integrate the peak areas of all MRM ion pairs using Agilent's Mass Hunter Quantitative Analysis software to obtain the absolute response of SIS and NAT. Recall, that the NAT signal should co-elute with its SIS counterpart and exhibit a similar peak shape since they have identical physicochemical properties. A differing peak shape suggests partial or complete spectral overlap of the precursor and/or product ion with a nontarget ion.

Table 2
Example of a typical output from an interference determination experiment performed on the standard-flow UHPLC/MRM-MS platform. Shown are the relative ratios calculated from the observed absolute responses of SIS in buffer, SIS in plasma, and NAT in plasma for the top 5 MRM transitions from intercellular adhesion molecule 1 (peptide LLGIETPLPK). Based on the average relative ratios and variability, the y_8^+ MRM ion pair would be selected as the target transition (referred to as the quantifier). If two additional transitions are required for added peak confirmation, the y_6^+- and y_4^+-containing MRM ion pairs would be selected (referred to as the qualifier peaks). In the qualifier selection, the y_4^+ ion pair would be selected over the y_9^{2+} ion pair despite having a marginally lower average relative ratio, because it is below the 20 % CV cutoff

	Relative ratios			
Product ion	**SIS in buffer**	**SIS in plasma**	**NAT in plasma**	**Av. relative ratio (%CV)**
b_4^+	0.09	0.09	0.11	0.09 (16)
y_4^+	0.17	0.16	0.19	0.17 (3)
y_6^+	0.27	0.26	0.24	0.25 (6)
y_8^+	1.00	1.00	1.00	1.00 (0)
y_9^{2+}	0.13	0.13	0.28	0.18 (45)

(e) Calculate the relative ratio of each peptide's transitions for the SIS form in buffer, the SIS form in plasma, and the NAT form in plasma.

(f) Determine the average relative ratio of the peptides MRM ion pairs and the %CV between SIS in buffer, SIS in plasma, and NAT in plasma.

(g) Rank each peptide's transitions according to the average relative ratio, as long as the variability does not exceed 20 % (see Table 2).

While it is common to monitor three transitions per precursor (i.e., one quantifier and two qualifiers), we construct our final MRM methods with a single representative transition for each proteotypic peptide. The selected transition is the one that returns the highest average relative ratio and is interference-free, with a CV below 20 %. This approach has shown sufficient specificity and reliability for peptide quantitation of complex mixtures (14, 15, 23). Using the final MRM acquisition method in combination with the optimal LC gradient and system loading amounts, typical extracted ion chromatograms (XIC) traces for panels of CVD-related biomarkers measured on a standard-flow and nano-flow LC/MRM-MS platform are illustrated in Fig. 2.

3.6 Protein Quantitation

Peptide calibration curves contain a wealth of information that is needed for the determination of such performance metrics as the lower limit of quantitation (LLOQ), the dynamic range of the assay, and the plasma protein concentration. In our laboratory, we typically construct these curves using a series of standard samples that

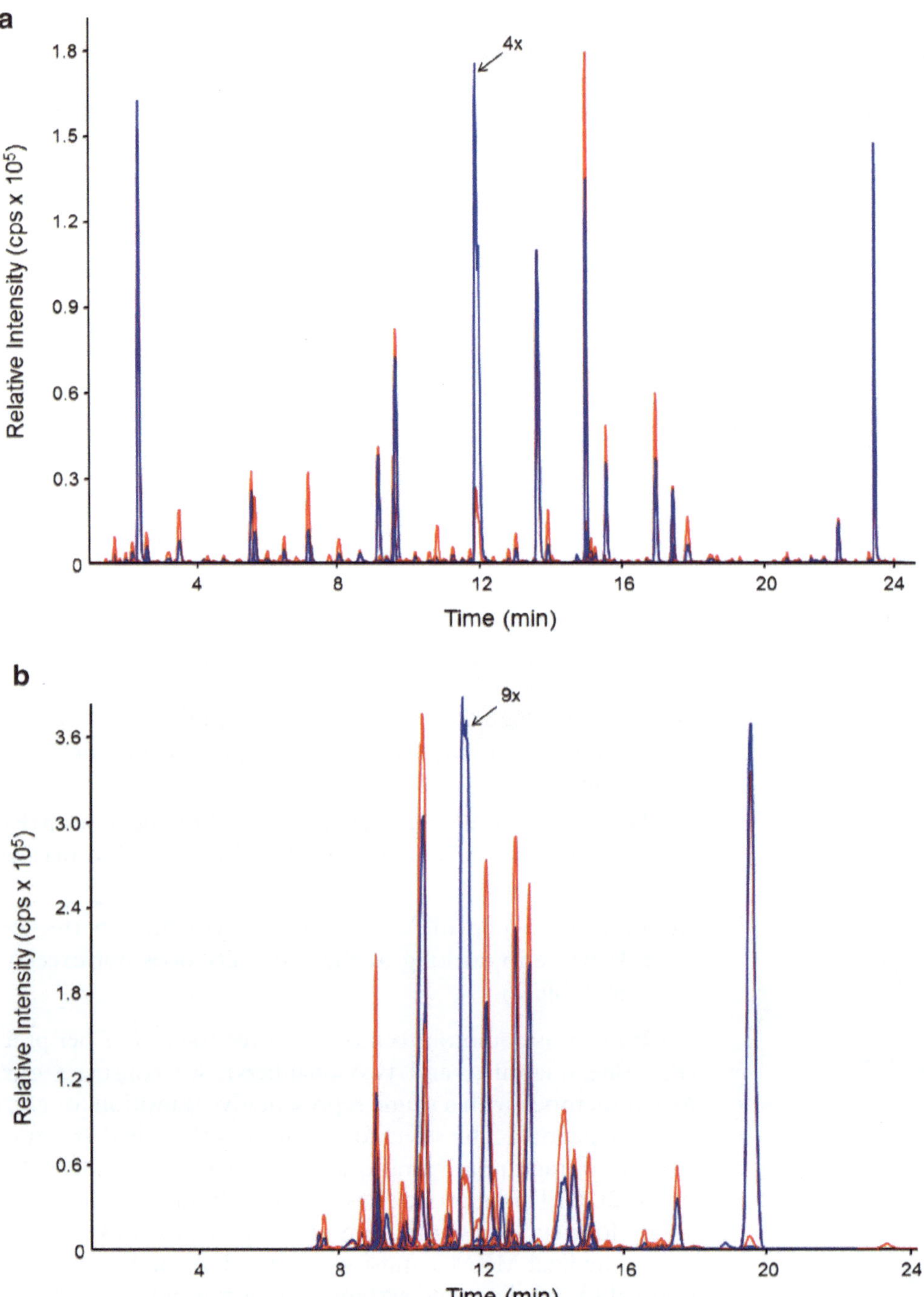

Fig. 2 XIC trace of a panel of MRM protein assays in a single 30 min LC/MRM-MS analysis. (**a**) XICs for 67 proteins (135 peptides, 270 scheduled ion pairs) from a human plasma digest measured with the standard-flow LC/MRM-MS platform. (**b**) XICs for 25 proteins (88 peptide, 176 transitions) from a human plasma digest measured on the nanoLC/MRM-MS platform. The signal intensity of albumin is reduced by a factor of 4 and 9 in parts (**a**) and (**b**), respectively. In both parts, the NAT peptides are displayed in *blue*, while the SIS peptides are in *red*. (**a**) is reprinted from ref. 15 with permission; (**b**) is reprinted from ref. 14, with permission

contain a variable concentrations of SIS peptides and a constant amount of plasma digest to keep the matrix effect constant, although we have also prepared samples using the reverse – variable NAT and constant SIS (*see* Subheading 3.5.3 and Kuzyk et al. (19) for protocols).

1. Prepare a series of digests containing a constant amount of NAT and a variable concentration of SIS peptides or the reverse, spanning a 3–4 order of magnitude range (see Table 1 for details from the platform comparison study).
2. Create a dynamic MRM-MS acquisition method for the SIS and NAT peptides using the top, interference-free, signal-producing ion in each case.
3. Use the LC parameters outlined in Subheading 3.3 and the optimal gradient determined in Subheading 3.5.2.
4. Load the standard samples onto the UHPLC or HPLC-Chip column for ≥3 replicate LC/MRM-MS analyses (*see* **Note 11**).
5. Manually inspect the integrated peaks in the Mass Hunter Quantitative Analysis software to ensure correct peak detection and accurate integration.
6. In performing the linear regression of the calibration curve, we use a $1/x$ or $1/x^2$ (x = concentration) weighting option to assist in covering a wide dynamic range.
7. The standard points within each sample concentration must be both precise (20 % CV) and accurate (80–120 %), as advised by the US FDA (24), to remain in the calibration curve. If the points for a given concentration fall outside these ranges, that concentration must be removed from the curve. In the end, a minimum of three concentrations must remain in order to construct a calibration curve.
8. The LLOQ is defined as the lowest point on the curve that has a CV below 20 % and an average accuracy in determining the expected concentration within 80–120 %. The LLOQ (in fmol/μL) can readily be converted to ng/mL by taking into account the weight of the entire processed protein sample (*see* **Note 13**).
9. The dynamic range is defined as the difference between the LLOQ and the ULOQ.
10. The concentration of each peptide target (in fmol/μL of plasma) is calculated based on the observed relative response and the linear regression equation as determined from the calibration curve. The determined concentration can also be converted to ng/mL by taking into account the molecular weight of the protein.

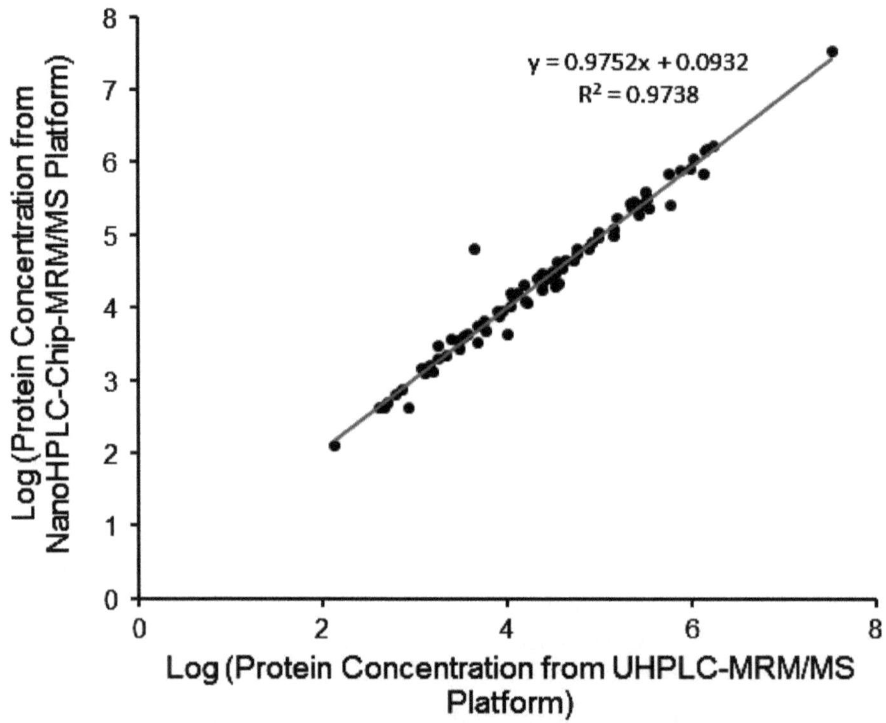

Fig. 3 Comparison of plasma protein concentrations measured by the standard-flow and nano-flow LC/MRM-MS platforms. Protein concentrations for an 81-peptide panel (corresponding to 48 CVD-related plasma proteins) determined from the two platforms are plotted against one another. Reprinted from ref. 14, with permission

In comparing the quantitative performance of a standard-flow UHPLC system to a nano-flow HPLC-Chip with a panel of 48 CVD-related plasma proteins (14), we found the standard-flow LC/MRM-MS platform to return lower LLOQs (average LLOQ: 2.6 μg/mL, standard-flow; 26.5, nano-flow) and a superior linear dynamic range (an average of one order of magnitude wider). In fact, 14 % more peptides could be quantified at the lower limit of the concentration range investigated in the standard-flow platform. The lowest quantifiable concentration in both platforms came from peptide TLAFPLTIR from the endothelial cell protein C receptor protein with a LLOQ of 4.3 ng/mL in the standard-flow and 43 ng/mL in the nano-flow. There was one peptide (SSPVVIDASTAIDAPSNLR from fibrinonectin), however, that was only quantifiable in the nano-flow platform (LLOQ: 232 fmol; determined plasma protein concentration: 1,061 fmol/μL or 275,410 ng/mL). This demonstrates the advantage of a nano-flow system in certain cases. Nonetheless, if sufficient sample is available, a standard-flow LC/MRM-MS platform is preferred. As expected, the plasma protein concentrations determined from the two platforms were nearly identical (see Fig. 3), with a coefficient of determination of 0.97 and a

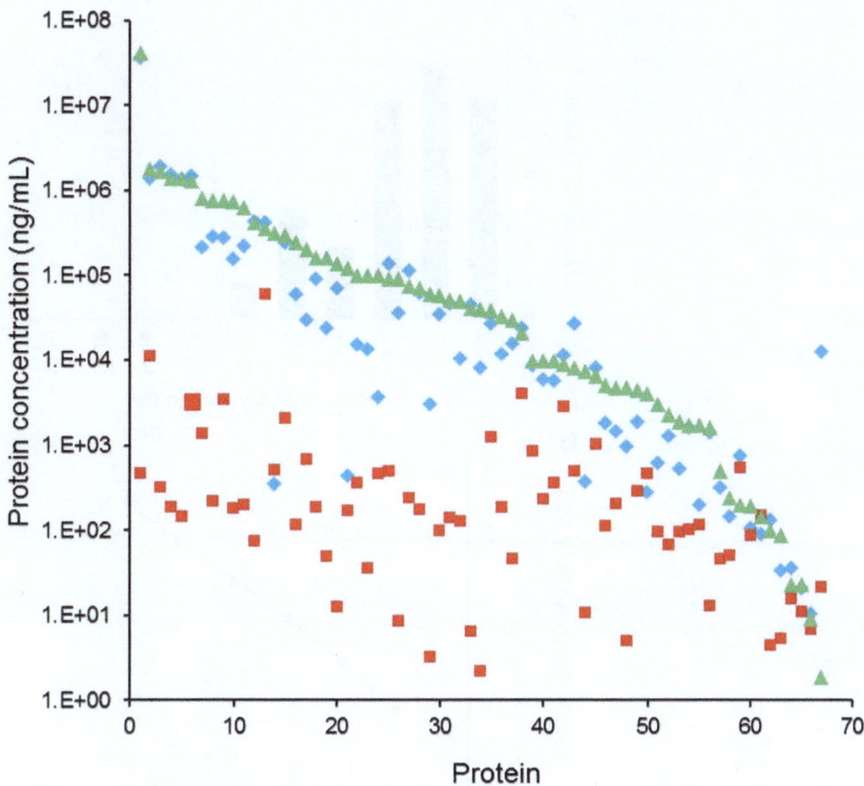

Fig. 4 Determined plasma protein concentrations of the 67 CVD-related protein biomarkers ordered from highest to lowest abundance (*blue diamonds*) along with the LLOQ of each MRM assay (*red squares*). These calculated values are compared to the mean reported plasma protein concentrations (*green triangles*). Reprinted from ref. 15, with permission

linear regression slope of 0.98. This indicates the ability of the two platforms to provide equal quantitative results.

Using the standard-flow LC/MRM-MS platform, together with the aforementioned strategies, we were able to quantitate 67 CVD-associated plasma proteins ranging from albumin (determined protein concentration: 36 mg/mL) to insulin-like growth factor binding protein 1 (determined protein concentration: 10 ng/mL) in a 30 min multiplexed analysis (see Fig. 4) (15). The seven order of magnitude range in protein concentrations observed in this study spans roughly 70 % of the range of concentrations inherent in plasma and is at least two orders of magnitude deeper than that previously attained with non-prefractionated human plasma (22). In addition, 25 of the 67 assays had LLOQs between 2 and 100 ng/mL, which brings these assays into the upper range of detection capabilities of the ELISAs (3) and rivals the level of sensitivity expected from MRM protein assays that use depletion (9, 10) or enrichment (12, 13).

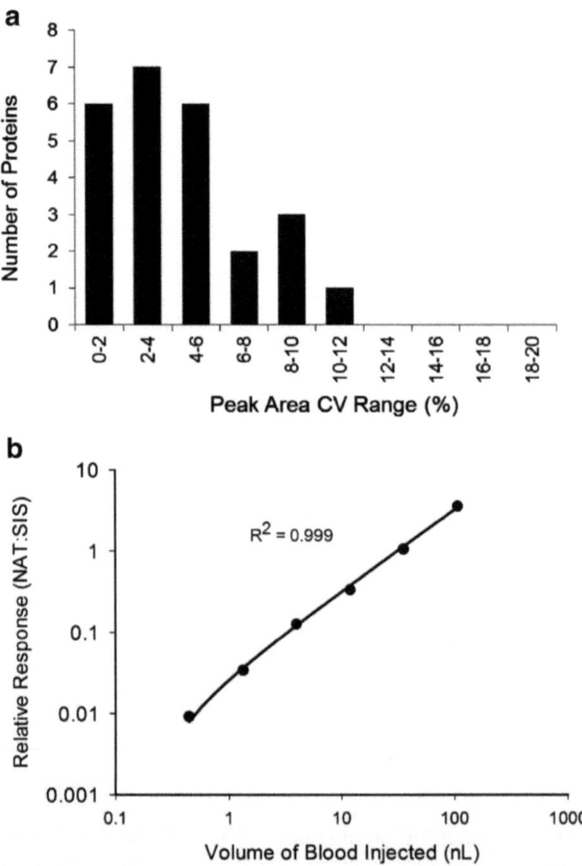

Fig. 5 Quantitation of proteins in DBS samples. (**a**) CV values for the normalized peak area ratios (NAT:SIS) for 25 targeted proteins. (**b**) Calibration curve for fibrinogen gamma chain (target peptide: YEASILTHDSSIR) generated from six samples containing a constant amount of SIS peptide and a variable amount of whole blood spotted on the DBS collection card

A subset containing 36 of the 67 CVD-associated plasma proteins from the previous application were also targeted in DBS samples using the standard-flow LC/MRM-MS platform. A total of 25 proteins (70 %) from this list were readily observed in DBS and could be quantitated. Both high molecular weight proteins (e.g., apolipoprotein B-100, 513 kDa) and low molecular weight proteins (e.g., apolipoprotein C-III, 9 kDa) were successfully extracted from the DBS collection cards. Replicate sample processing and MRM analysis of the 25 proteins resulted in average CV values for the normalized peak area ratios (NAT:SIS) of only 4.4 %, which demonstrates the high reproducibility of our DBS sample processing method. Calibration curves were generated by serial dilution of whole blood prior to spotting and revealed a linear dynamic range of over 200 for some proteins (see Fig. 5). The dynamic range for the entire assay was more than three orders of magnitude as the concentration of the highest and lowest abundance proteins in whole blood were determined to be 49 mg/mL for albumin and 11 μg/mL for coagulation factor XII HC.

3.7 Conclusions and Significance

While numerous strategies for discovering disease-related biomarkers exist, the bottleneck in driving these discovered markers toward clinical use remains in the verification and validation phases of the biomarker development pipeline (25). Validation in particular requires the analysis of an exceedingly large number of patient samples to be certain that the target molecule is a true biomarker of the disease. With this goal in mind, research development efforts strive to achieve accurate and absolute protein quantitation in a high-throughput and multiplexed manner. An emerging approach that can accomplish this uses MRM-MS in combination with SIS peptides. Front-end immunoaffinity enrichment or depletion is a common add-on with this approach to increase the depth of coverage. Immunoprecipitation, however, increases the cost, reduces the overall assay throughput, and may compromise the intra-sample reproducibility. Therefore, we compared the analytical performance of a standard-flow UHPLC and a nano-flow HPLC-Chip system interfaced to Agilent's latest-generation triple quadrupole mass spectrometer (the 6490) in a technological comparison study for quantitating proteins in non-prefractionated human plasma. This study revealed the standard-flow platform to be more sensitive and robust in quantitating 48 putative CVD plasma biomarkers (14). The nano-flow platform still, however, has merit for certain peptides (e.g., SSPVVIDASTAIDAPSNLR from fibrinonectin) and under sample-limited conditions (ca. 14 nL of raw plasma injected/analysis). The methods for analyzing plasma samples on both platforms were therefore described in this chapter.

We have applied the desired standard-flow platform to the analysis of human plasma and human DBS samples in several applications. In one application, we were able to quantitate 67 plasma proteins covering the top seven order of magnitude range of protein concentration, with seven of those present in the 10–100 ng/mL range (15). The low LLOQs obtained are comparable to those reported from prefractionation-based methods and bring this MRM-MS assay into the range where it can detect tissue leakage proteins, which are strong indicators of disease, being secreted into the bloodstream during only necrosis, apoptosis, or hemolysis. In a second application, the standard-flow platform demonstrated the ability to quantitate 25 putative CVD protein biomarkers in DBS samples. With minor changes to the protein extraction procedure, the developed method may also be applicable to other dried biological samples including plasma, serum, saliva, and urine.

Overall, the protocols described in this chapter are inexpensive, in terms of time and reagent costs, and satisfy the throughput and multiplexing requirements for the verification and validation stages of a biomarker project. It is our hope that these protocols will become adopted in large-scale studies to expedite the verification and validation of discovered candidate protein biomarkers.

4 Notes

1. Prior to spotting, the DBS collection card should be bent slightly so that the back of the sample area is not in contact with any surface. This prevents losses from blood that soaks through the card and reduces the possibility of contaminating surfaces with hazardous blood samples. The sample is expelled from the pipette tip with the hanging droplet brought into contact with the DBS collection card.
2. Alternatively, the center of the DBS spot (3 or 6 mm diameter) can be removed using a single-hole punch or a disposable biopsy punch.
3. We use sodium deoxycholate for protein denaturation over alternative chaotropes (e.g., urea) and detergents (e.g., sodium dodecyl sulfate) for two main reasons. First, it is acid insoluble and can thus be readily removed following tryptic proteolytic digestion through acid quenching. This expedites sample cleanup and minimizes the concern of ion suppression during LC-MS from the surfactant itself. Second, in a side-by-side comparison study of common denaturants, sodium deoxycholate was observed to give enhanced digestion efficiency (26), which indicated its power in destabilizing the non-covalent interactions for improved access to the active site of the trypsin enzyme.
4. We synthesize and purify C-terminal isotopically coded standard peptides in-house with ($^{13}C_6$, $^{15}N_2$)- or ($^{13}C_6$)-lysine and ($^{13}C_6$, $^{15}N_4$)- or ($^{13}C_6$)-arginine, according to a previously documented protocol (19). These provide a minimum mass difference between the NAT and SIS peptides of 6 Da, which is sufficient to prevent overlapping precursor and product ion isotopic distributions. Further, the ^{13}C- and ^{15}N-labeled analogues chromatographically co-elute with its unlabeled counterpart, unlike the ^{2}H isotope, which aids peak identification. The SIS peptides are stored at −80 °C and only aliquots removed for addition to the tryptic digest.
5. Remove the supernatant for solid-phase extraction by carefully withdrawing of the solution along one side of the Eppendorf tube to prevent disrupting the sodium deoxycholate precipitate.
6. With a final reconstituted protein concentration of 1 μg/μL, a 10 and 1 μL injection of digest onto the standard-flow and nano-flow LC/MRM-MS platforms equates to 143 and

14.3 nL of raw plasma for analysis, respectively (based on an initial plasma protein concentration of 70 mg/mL).

7. We use the ProtID-Chip-150 which contains, in addition to the HPLC separation column, an enrichment column (40 nL, Zorbax 300SB-C_{18}, 5 μm particles) for sample concentration and a nanospray emitter (50 μm i.d.) for peptide ionization.
8. While the gradient optimization on the standard-flow LC system can be conducted for just SIS in buffer, both buffer and plasma must be examined in the nano-flow system. In our experience, we found the retention times of SIS in buffer to consistently differ from SIS in plasma by roughly 1 min.
9. Since the SIS peptides separate, ionize, and fragment identically to their NAT, matching MRM acquisition parameters and retention times can be used for both peptide forms. The only difference is in the actual precursor and product ion *m/z* values for the NAT peptides, which can be readily determined from the Peptide Sequence Fragmentation Modelling tool of the Molecular Weight Calculator program. This program is a Windows-based freeware utility that was developed by Matthew Monroe at the Pacific Northwest National Laboratory and is available for download at http://omics.pnl.gov/software/MWCalculator.php.
10. We strive to analyze our target peptides within a 30 min chromatographic run time; however, if the number of targets become excessive, it is reasonable to lengthen the gradient and/or diminish the MRM detection window for improved chromatographic resolution.
11. To minimize sample carryover, we recommend analyzing the standard samples in order of increasing concentration, with blank solvent injections run between each concentration.
12. While it can be argued that 50 μg on the standard-flow column may be more desirable (e.g., peptide 5 and 10 in Fig. 2a), this is not typical and can lead to excess sample deposition in the capillary inlet and ion funnel, which hinders ion transmission to the Q1 mass analyzer.
13. We use the UniProtKB/Swissprot database on the Expasy Web server at http://www.uniprot.org/ to locate the molecular weight of the target proteins.

Acknowledgments

We wish to thank Genome Canada, Genome BC, and the Western Economic Diversification of Canada for providing funding to the UVic Genome BC Proteomics Centre to perform this research.

References

1. Anderson L (2005) Candidate-based proteomics in the search for biomarkers of cardiovascular disease. J Physiol 563:23–60
2. Anderson NL (2010) The clinical plasma proteome: a survey of clinical assays for proteins in plasma and serum. Clin Chem 56:177–185
3. Anderson NL, Anderson NG (2002) The human plasma proteome: history, character, and diagnostic prospects. Mol Cell Proteomics 1:845–867
4. Chace DH, Kalas TA (2005) A biochemical perspective on the use of tandem mass spectrometry for newborn screening and clinical testing. Clin Biochem 38:296–309
5. Want EJ, Cravatt BF, Siuzdak G (2005) The expanding role of mass spectrometry in metabolite profiling and characterization. Chembiochem 6:1941–1951
6. Addona TA, Abbatiello SE, Schilling B, Skates SJ, Mani DR, Bunk DM, Spiegelman CH et al (2009) Multi-site assessment of the precision and reproducibility of multiple reaction monitoring-based measurements of proteins in plasma. Nat Biotechnol 27:633–641
7. Parker CE, Pearson TW, Anderson NL, Borchers CH (2010) Mass spectrometry based clinical proteomics, a review and prospective. Analyst 135:1830–1838
8. Huttenhain R, Malmstrom J, Picotti P, Aebersold R (2009) Perspectives of targeted mass spectrometry for protein biomarker verification. Curr Opin Chem Biol 13:518–525
9. Keshishian H, Addona T, Burgess M, Kuhn E, Carr SA (2007) Quantitative, multiplexed assays for low abundance proteins in plasma by targeted mass spectrometry and stable isotope dilution. Mol Cell Proteomics 6:2212–2229
10. Keshishian H, Addona T, Burgess M, Mani DR, Shi X, Kuhn E, Sabatine MS, Gerszten RE, Carr SA (2009) Quantification of cardiovascular biomarkers in patient plasma by targeted mass spectrometry and stable isotope dilution. Mol Cell Proteomics 8:2339–2349
11. Zolotarjova N, Mrozinski P, Chen H, Martosella J (2008) Combination of affinity depletion of abundant proteins and reverse-phase fractionation in proteomic analysis of human plasma/serum. J Chromatogr A 1189:332–338
12. Whiteaker JR, Zhao L, Anderson L, Paulovich AG (2010) An automated and multiplexed method for high throughput peptide immunoaffinity enrichment and multiple reaction monitoring mass spectrometry-based quantification of protein biomarkers. Mol Cell Proteomics 9:184–196
13. Whiteaker JR, Zhao L, Abbatiello SE, Burgess M, Kuhn E, Lin C, Pope M, Razavi M, Anderson NL, Pearson TW, Carr SA, Paulovich AG (2011) Evaluation of large scale quantitative proteomic assay development using peptide affinity-based mass spectrometry. Mol Cell Proteomics 10:M110.005645
14. Percy AJ, Chambers AG, Yang J, Domanski D, Borchers CH (2012) Anal Bioanal Chem 404: 1089–1101
15. Domanski D, Percy AJ, Yang J, Chambers AG, Hill JS, Cohen Freue GV, Borchers CH (2012) MRM-based multiplexed quantitation of 67 putative cardiovascular disease biomarkers in human plasma. Proteomics 12:1222–1243
16. Chambers AG, Percy AJ, Yang J, Camenzind AG, Borchers CH (2012) Mol Cell Proteomics in press
17. Li W, Tse FLS (2010) Dried blood spot sampling in combination with LC-MS/MS for quantitative analysis of small molecules. Biomed Chromatogr 24:49–65
18. Keevil BG (2011) The analysis of dried blood spot samples using liquid chromatography tandem mass spectrometry. Clin Biochem 44:110–118
19. Kuzyk MA, Parker CE, Borchers CH (2012) In: Backvall H (ed) Methods in molecular biology. Humana Press, still in press
20. Anderson L, Hunter CL (2006) Quantitative mass spectrometric multiple reaction monitoring assays for major plasma proteins. Mol Cell Proteomics 5:573–588
21. Unwin RD, Griffiths JR, Leverentz MK, Grallert A, Hagan IM, Whetton AD (2005) Multiple reaction monitoring to identify sites of protein phosphorylation with high sensitivity. Mol Cell Proteomics 4:1134–1144
22. Kuzyk MA, Smith D, Yang J, Cross TJ, Jackson AM, Hardie DB, Anderson NL, Borchers CH (2009) Multiple reaction monitoring-based, multiplexed, absolute quantitation of 45 proteins in human plasma. Mol Cell Proteomics 8:1860–1877
23. Gerber SA, Rush J, Stemman O, Kirschner MW, Gygi SP (2003) Absolute quantification of proteins and phosphoproteins from cell lysates by tandem MS. Proc Natl Acad Sci USA 100:6940–6945
24. US Department of Health and Human Services, Food and Drug Administration. Guidance for Industry Bioanalytical Method Validation (2001) http://www.fda.gov/downloads/Drugs/GuidanceComplianceRegulatoryInformation/Guidances/ucm070107.pdf

25. Surinova S, Schiess R, Hüttenhain R, Cerciello F, Wollscheid B, Aebersold R (2011) On the development of plasma protein biomarkers. J Proteome Res 10:5–16
26. Proc JL, Kuzyk MA, Hardie DB, Yang J, Smith DS, Jackson AM, Parker CE, Borchers CH (2010) A quantitative study of the effects of chaotropic agents, surfactants, and solvents on the digestion efficiency of human plasma proteins by trypsin. J Proteome Res 9: 5422–5437

Chapter 14

Multiple Reaction Monitoring (MRM) of Plasma Proteins in Cardiovascular Proteomics

Verónica M. Dardé, Maria G. Barderas, and Fernando Vivanco

Abstract

Different methodologies have been used through years to discover new potential biomarkers related with cardiovascular risk. The conventional proteomic strategy involves a discovery phase that requires the use of mass spectrometry (MS) and a validation phase, usually on an alternative platform such as immunoassays that can be further implemented in clinical practice. This approach is suitable for a single biomarker, but when large panels of biomarkers must be validated, the process becomes inefficient and costly. Therefore, it is essential to find an alternative methodology to perform the biomarker discovery, validation, and quantification. The skills provided by quantitative MS turn it into an extremely attractive alternative to antibody-based technologies. Although it has been traditionally used for quantification of small molecules in clinical chemistry, MRM is now emerging as an alternative to traditional immunoassays for candidate protein biomarker validation.

Key words Biomarker discovery, Human plasma, Cardiovascular proteomics, Multiple reaction monitoring, Candidate biomarker validation

1 Introduction

Due to the complex and multifactorial basis of cardiovascular diseases, it is essential to develop biomarkers for early detection, diagnosis, prognosis, therapy monitoring, etc. Differential proteomics provides a powerful tool to study global changes in protein expression in cardiovascular diseases, and it has enabled the identification of new potential biomarkers implicated in the origin and development of cardiovascular pathologies in the past years (1–7). However, these candidate biomarkers must be tested and validated in larger sample cohorts before they can be applied in clinical practice (8). To date, most validation procedures were based on immunoaffinity methods that can be very sensitive, but they must be standardized before they can be employed in routine analyses. Besides, highly specific antibodies must be available and the development of such techniques is expensive and time-consuming.

Fernando Vivanco (ed.), *Vascular Proteomics: Methods and Protocols*, Methods in Molecular Biology, vol. 1000, DOI 10.1007/978-1-62703-405-0_14,

It is well established that mass spectrometry is a reliable methodology for quantification of small molecules and targeted MS has been widely used in analytical and clinical chemistry (9–13). Lately, multiple reaction monitoring (MRM) using a triple quadrupole mass spectrometer has been proved to be also very useful for quantification of proteins. This strategy offers absolute structural specificity for the selected analyte, and it increases the sensitivity for peptides (as compared with unbiased MS analysis) (14). Moreover, it can provide quantitative information based on chromatographic peak integration that can be either relative or absolute quantification of analyte concentration by adding known amounts of stable isotope-labeled standards. And since it is possible to monitor several analytes in one single run (15), MRM is becoming a valuable alternative to traditional immunoassays for candidate biomarker validation, and it has been successfully applied in complex protein mixtures, including plasma (16–18).

There are mainly two strategies to develop MRM assays for peptide analysis: compound optimization by infusion of a synthetic standard peptide or MRM assay development without any synthetic standard molecules. In the first strategy, every parameter affecting the signal quality can be optimized to generate the finest acquisition method. The second strategy presents a higher throughput, but optimization is only possible for certain parameters. In this workflow a candidate list of MRM transitions is first generated, using previous MS/MS data or using peptide sequence to calculate theoretical precursor masses and predicted fragment data (Fig. 1). Then, this candidate list of transitions is tested in a sample containing the target peptides. To perform this test, it is strongly recommended to use a hybrid triple quadrupole-linear ion trap (3Q-LIT) mass spectrometer (19, 20), since it allows to perform a kind of information-dependent experiment known as MIDAS (MRM-initiated detection and sequencing) (21), where enhanced product ion (EPI) scans are triggered by a specific SRM peak, linking a MS/MS spectrum to a certain SRM signal. Resulting MS/MS data must confirm that the precursor ions match the predicted peptide sequences, and they can be also used to optimize the initial transition list, by creating a new candidate list of transitions containing the most abundant fragments in MS/MS spectra. This procedure is repeated until an optimal final transition list is generated, containing those transitions that can be unambiguously related to a certain peptide and that can be used for quantification.

2 Materials

2.1 Equipment

1. Shaker (Thermo Espresso Personal Microcentrifuge, Thermo Fisher Scientific Inc., MA, USA).
2. Vortex mixer (Stuart Scientific SA8 vortex mixer, Barloworld Scientific Ltd., UK).

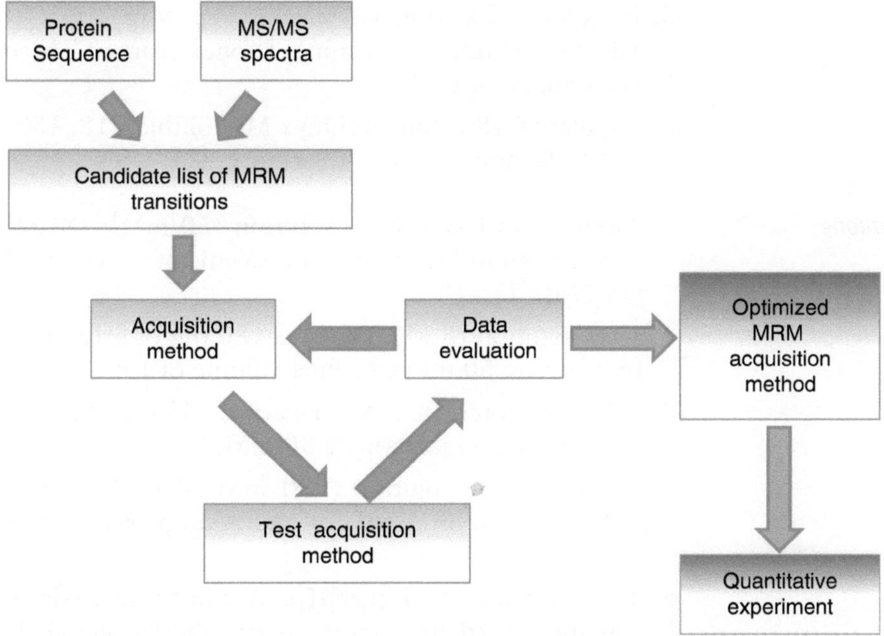

Fig. 1 Development of a MRM assay without any synthetic standard molecules. Starting from the theoretical sequence of the protein of interest and/or MS/MS data for that protein, theoretical transitions can be calculated (using different software packages). This candidate list is then tested in a sample containing the target peptides. Evaluation of resulting data can generate a new candidate list of transitions, and this iterative method can be repeated until an optimal reliable transition list that can be used for quantification is obtained

3. Thermomixer comfort (Eppendorf AG. Hamburg, Germany).
4. Convection oven.
5. SpeedVac (Savant SPD111V SpeedVac Concentrator, Thermo Fisher Scientific Inc., MA, USA).
6. NanoHPLC system with autosampler (TEMPO nano LC system, AB Sciex, CA, USA).
7. Triple quadrupole-linear ion trap mass spectrometer (4000 QTRAP LC/MS/MS System, AB Sciex, CA, USA).
8. Analyst Software version 1.5.1 (AB Sciex, CA, USA).
9. MRM Pilot software version 2.1 (AB Sciex, CA, USA).

2.2 Reagents

1. Dithiothreitol (DTT), iodoacetamide, and formic acid, Sigma-Aldrich Corporation (LLC., MO, USA).
2. Ammonium bicarbonate, acetonitrile LC/MS grade and water LC/MS grade.
3. Sequencing grade modified porcine trypsin purchased from Promega Corporation (WI, USA).
4. PepClean spin columns purchased from Pierce (part of Thermo Fisher Scientific Inc., IL, USA).

5. Precolumn Cartridges (Acclaim Pep Map 100 C18, 5 μm, 100 Å; 300 μm i.d. × 5 mm), Dionex (part of Thermo Fisher Scientific, CA, USA).
6. Capillary C18 columns (Onyx Monolithic C18, 150 × 0.1 mm I.D.), Phenomenex Inc. (Torrance, CA, USA).

2.3 Solutions

1. Ammonium bicarbonate solution (50 mM): Dissolve 99 mg of ammonium bicarbonate in 25 mL of water LC/MS grade (*see* **Note 1**).
2. Urea solution (8 M): Dissolve 480 mg of urea in ammonium bicarbonate 50 mM to a final volume of 1 mL.
3. DTT solution (100 mM): Dissolve 153 mg of DTT in 10 mL of ammonium bicarbonate 50 mM.
4. Iodoacetamide solution (550 mM): Dissolve 101.75 mg of iodoacetamide in 1 mL of ammonium bicarbonate 50 mM (*see* **Note 1**).
5. Trypsin solution (1 μg/μL): Reconstitute 1 vial of trypsin (20 μg) in 20 μL of the specific buffer supplied with the enzyme.
6. PepClean reconstitution and elution solution: Add 40 μL of formic acid to 9.5 mL of acetonitrile LC/MS and fill up to 10 mL with water LC/MS grade.
7. PepClean solution 1: Mix 200 μL of acetonitrile LC/MS grade with 50 μL of formic acid and fill up to 10 mL with water LC/MS grade.
8. PepClean solution 2: Mix 200 μL of acetonitrile LC/MS grade with 10 μL of formic acid and fill up to 10 mL with water LC/MS grade.
9. Tryptic digest reconstitution solution: Mix 2 mL of acetonitrile LC/MS grade with 2 mL of formic acid and fill up to 100 mL with water LC/MS grade.
10. Phase A: Add 100 μL of formic acid to 2 mL of acetonitrile LC/MS and fill up to 100 mL with water LC/MS grade.
11. Phase B: Add 100 μL of formic acid to 98 mL of acetonitrile LC/MS and fill up to 100 mL with water LC/MS grade.

3 Methods

3.1 Tryptic Digestion

Tryptic digestion is typically used in proteomics before MS analysis, since trypsin cleaves peptides in a predictable manner, at the carboxyl side of the amino acids lysine or arginine (except if either of them is followed by a proline). Besides, this digestion generates peptides with an average size of ten amino acids, in a mass range suitable for most mass spectrometers. In order to facilitate enzyme access to its substrate and to optimize peptide yield after digestion, it is important to denature proteins, reduce cysteines, and alkylate resulting sulfhydryl groups to prevent disulfide bond formation. Trypsin is active at basic pH so the reaction buffer must be ammonium bicarbonate or similar (*see* **Note 2**).

3.1.1 Sample Denaturation, Reduction, and Alkylation

Start with a volume of sample below 10 μL (containing 20–100 μg of protein):

1. Add 10 μL of urea 8 M to the sample. Adjust the sample volume to 20 μL by adding water LC/MS grade.
2. Add 2 μL of DTT 100 mM and incubate 30 min at 37 °C in a thermomixer (*see* **Note 3**).
3. Add 2.2 μL of iodoacetamide 550 mM and incubate 20 min at room temperature in the dark.

3.1.2 Digestion Reaction

1. Add 60 μL of ammonium bicarbonate 50 mM and 15 μL of acetonitrile LC/MS grade (*see* **Note 4**).
2. Add 1 μL of trypsin (1 μg/μL) to every 50 μg of protein sample.
3. Incubate at 37 °C overnight in a convection oven.
4. Stop reaction by adding 2 μL of formic acid.

3.1.3 Sample Digests Cleaning

1. Dry samples in a SpeedVac concentrator (*see* **Note 5**).
2. Resuspend them in 100 μL of PepClean solution 1.
3. Place as many PepClean spin columns as digested samples exist in clean 1.5 mL tubes.
4. Add 300 μL of PepClean reconstitution and elution solution to the top of the PepClean spin columns. Centrifuge and discard flow-through.
5. Add 300 μL of PepClean solution 1 to the top of the PepClean spin columns. Centrifuge and discard flow-through.
6. Repeat **step 5** once.
7. Apply each sample to the top of one PepClean spin column and place it into the tube that contained the corresponding sample. Centrifuge and recover the flow-through.
8. Repeat **step** 7 once.
9. Add 300 μL of PepClean solution 2 to the top of the PepClean spin columns. Centrifuge and discard flow-through.
10. Repeat **step 9** twice.
11. Place every PepClean spin column in a new clean 1.5 mL tube.
12. Add 50 μL of PepClean reconstitution and elution solution to the top of the PepClean spin columns. Centrifuge and recover the flow-through.

13. Repeat **step 12** and combine both flow-through fractions.
14. Dry samples in a SpeedVac concentrator.

3.2 MRM Analysis

3.2.1 MRM Assay Development Without Any Synthetic Standard Molecules

Theoretical SRM transitions can be designed using MRMPilot software (AB Sciex), introducing the theoretical sequence of the protein, indicating the parameters for the in silico digestion and the number of transitions to be monitored for each peptide (usually two or three transitions). Previous MS/MS data can also be imported if available into this software, improving its prediction capacity. A MIDAS acquisition method is then generated including all the theoretical transitions considered (*see* **Note 6**). Then, a sample containing a mixture of all the proteins of interest previously digested is injected and analyzed in the 4000QTrap, following these steps:

1. Resuspend the sample in tryptic digest reconstitution solution.
2. Program autosampler module to pick an aliquot of sample and to load it onto a μ-Precolumn Cartridge to preconcentrate and further desalt the sample.
3. Program the nanoLC system to perform a gradient of Phase A and Phase B to transfer the peptides from the μ-Precolumn Cartridge to the capillary C18 column and to separate them along the chromatographic run. Peptides will then be eluted at a flow rate of 900 nL/min in the following continuous acetonitrile gradient: 2–15 % B for 2 min, 15–30 % B for 18 min, 30–50 % B for 5 min, 50–90 % B for 2 min, and finally 90 % B for 3 min. Column is then regenerated with 2 % B for 15 additional minutes (*see* **Note 7**).
4. Set the mass spectrometer to operate in positive ion mode with ion spray voltage of 2,800 V and a nanoflow interface heater temperature of 150 °C. Source gas 1 and curtain gas should be set to 20 and 20 psi, using nitrogen as both curtain and collision gas. Acquire MIDAS data.
5. Evaluate MIDAS data. Select transitions when the two or three coeluting peaks (corresponding to the two or three transitions monitored for the same peptide) have a signal-to-noise ratio over 5 and the MS/MS data match the theoretical spectrum for that peptide. Collision energy can also be optimized to obtain maximum transmission efficiency and sensitivity for each transition.

This protocol can be repeated until an optimal transition list is generated. Then, a final MRM acquisition method (without the LIT scan modes, to maximize sensitivity) is generated.

3.2.2 Quantitative Data Acquisition

Once the final transition list is ready, it is possible to perform the same MRM analysis in many different samples (Fig. 2). Although

Final transition list

Protein	Accession code	Peptide Sequence	Q1	Q3	CE	Dwell time	Charge state	Fragment Ion
Apolipoprotein A-I	APOA1_HUMAN	VSFLSALEEYTK	693,86	853,43	36	50	+2	y7
			693,86	940,46	36	50	+2	y8
			693,86	1053,55	36	50	+2	y9
Apolipoprotein E	APOE_HUMAN	LGPLVEQGR	484,78	588,31	26	50	+2	y5
			484,78	701,39	26	50	+2	y6
			484,78	798,45	26	50	+2	y7
Serum paraoxonase/aryle sterase 1	PON1_HUMAN	IFFYDSENPPASEVLR	942,46	868,49	46,47	50	+2	y8
			942,46	982,53	46,47	50	+2	y9
			942,46	1198,61	46	50	+2	y11
Alpha-2-HS-glycoprotein	FETUA_HUMAN	HTLNQIDEVK	598,82	845,44	31	50	+2	y7
			598,82	958,52	31	50	+2	y8
			598,82	1059,57	31	50	+2	y9
Complement C3	CO3_HUMAN	VQLSNDFDEYIMAIEQTIK	753,04	802,47	41,65	50	+3	y7
			753,04	933,51	42	50	+3	y8
			753,04	1046,59	42	50	+3	y9
Vitamin D-binding protein	VTDB_HUMAN	HLSLLTTLSNR	627,86	804,46	33	50	+2	y7
			627,86	917,54	33	50	+2	y8
			627,86	1004,5	33	50	+2	y9

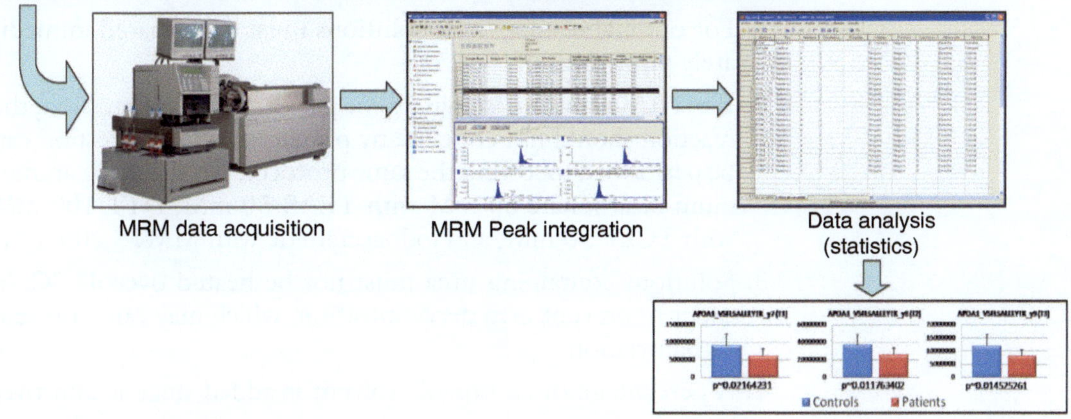

MRM data acquisition

MRM Peak integration

Data analysis (statistics)

Controls Patients

Fig. 2 Schematic representation of a MRM quantification experiment. Transitions in the final list are monitored in a group of samples. After data acquisition, SRM peak areas are integrated to calculate peptide abundances in every chromatographic run, and these data are then used to perform subsequent statistical analysis

the objective here is not an absolute quantification, some controls should be included in the list of transitions. These can include fragments of an external protein (not present in the natural samples) that would have been previously added as an external standard (a fixed amount) to every sample. The list should also contain transitions of at least one endogenous protein that is not differentially expressed in either group (as an internal standard). It is strongly recommended to perform two or three technical replicates for each digested sample (reconstituted in the appropriate volume of tryptic digest reconstitution solution), following these steps in every replicate (*see* **Note 8**):

1. Program Autosampler module to pick an aliquot of sample and to load it onto a μ-Precolumn Cartridge to preconcentrate and further desalt the sample.
2. Program the nanoLC system to perform a gradient of Phase A and Phase B as described in Subheading 3.2.1, **step 3**.
3. Set the mass spectrometer to operate in positive ion mode with the same parameters described before and acquire MRM data.

4. Calculate abundances on the basis of peak areas after integration of each SRM peak in every chromatographic run, using IntelliQuan algorithm included in Analyst software.

   Then means and standard deviations of abundances for each SRM peak in every sample can be calculated, and further statistical analysis can be applied to these data (Fig. 2).

4 Notes

1. For optimal results, these solutions must be prepared immediately prior to use.
2. If mTRAQ labeling is going to be performed after digestion, the reaction should not contain any primary amines. Digestion can be performed following the same protocol, but replacing ammonium bicarbonate 50 mM with TEAB 50 mM, DTT 100 mM with TCEP 50 mM, and iodoacetamide with MMTS 200 mM.
3. Solutions containing urea must not be heated over 37 °C, in order to prevent urea decomposition, which may cause protein carbamylation.
4. A percentage of an organic solvent is added since it improves trypsin activity.
5. This step is very important when acetonitrile is added to the digestion mixture, in order to evaporate any residual acetonitrile before cleaning, since PepClean system is based on a C18 resin.
6. Dwell times should be adjusted to optimize sensitivity vs. cycle times. To obtain reliable quantifications after peak integration, cycle time should not exceed 1/8 of peak width.
7. Gradient conditions can be modified depending on the sample complexity and the number of transitions to be monitored. However, we recommend a flow rate near 1 mL/min when using monolithic capillary columns.
8. It is recommended to perform one blank run between replicates and two blank runs between different samples.

Acknowledgments

This work was supported by grants from FIS PI11/01401 and FIS PI08/0970 and by SESCAM.

References

1. McGregor E, Dunn MJ (2003) Proteomics of heart disease. Hum Mol Genet 12(Spec No. 2):R135–R144
2. Van Eyk JE, Dunn MJ (2003) Proteomic and genomic analysis of cardiovascular disease. Wiley, Chichester
3. Mayr M, Mayr U, Chung YL, Yin X, Griffiths JR, Xu Q (2004) Vascular proteomics: linking proteomic and metabolomic changes. Proteomics 4:3751–3761
4. Marian AJ, Nambi V (2004) Biomarkers of cardiac disease. Expert Rev Mol Diagn 4:805–820
5. Barderas MG, Tunon J, Darde VM, De la Cuesta F, Duran MC, Jiménez-Nácher JJ, Tarín N, López-Bescós L, Egido J, Vivanco F (2007) Circulating human monocytes in the acute coronary syndrome express a characteristic proteomic profile. J Proteome Res 6:876–886
6. Darde VM, De la Cuesta F, Gil-Dones F, Alvarez-Llamas G, Barderas MG, Vivanco F (2010) Analysis of the plasma proteome associated with acute coronary syndrome: does a permanent protein signature exist in the plasma of ACS patients? J Proteome Res 9: 4420–4432
7. Gil-Dones F, Martin-Rojas T, Lopez-Almodovar LF, de la Cuesta F, Darde VM, Alvarez-Llamas G, Juarez-Tosina R, Barroso G, Vivanco F, Padial LR, Barderas MG (2010) Valvular aortic stenosis: a proteomic insight. Clin Med Insights Cardiol 4:1–7
8. Rifai N, Gillette MA, Carr SA (2006) Protein biomarker discovery and validation: the long and uncertain path to clinical utility. Nat Biotechnol 24:971–983
9. Perchalski R, Yost R, Wilder B (1982) Structural elucidation of drug metabolites by triple-quadrupole mass spectrometry. Anal Chem 54:1466–1471
10. Tiller PR, Cunniff J, Land AP, Schwartz J, Jardine I, Wakefield M, Lopez L, Newton JF, Burton RD, Folk BM, Buhrman DL, Price P, Wu D (1997) Drug quantitation on a benchtop liquid chromatography-tandem mass spectrometry system. J Chromatogr A 771: 119–125
11. Lee MS, Kerns EH (1999) LC/MS applications in drug development. Mass Spectrom Rev 18:187–279
12. Tai SS, Bunk DM, White ET, Welch MJ (2004) Development and evaluation of a reference measurement procedure for the determination of total 3,3,5-triiodothyronine in human serum using isotope-dilution liquid chromatography-tandem mass spectrometry. Anal Chem 76:5092–5096
13. Sannino A, Bolzoni L, Bandini M (2004) Application of liquid chromatography with electrospray tandem mass spectrometry to the determination of a new generation of pesticides in processed fruits and vegetables. J Chromatogr A 1036:161–169
14. Keshishian H, Addona T, Burgess M, Kuhn E, Carr SA (2007) Quantitative, multiplexed assays for low abundance proteins in plasma by targeted mass spectrometry and stable isotope dilution. Mol Cell Proteomics 6:2212–2229
15. Barr DB, Barr JR, Maggio VL, Whitehead RD Jr, Sadowski MA, Whyatt RM, Needham LL (2002) A multi-analyte method for the quantification of contemporary pesticides in human serum and plasma using high-resolution mass spectrometry. J Chromatogr B Analyt Technol Biomed Life Sci 778:99–111
16. Keshishian H, Addona T, Burgess M, Mani DR, Shi X, Kuhn E, Sabatine MS, Gerszten RE, Carr SA (2009) Quantification of cardiovascular biomarkers in patient plasma by targeted mass spectrometry and stable isotope dilution. Mol Cell Proteomics 8:2339–2349
17. Kuhn E, Wu J, Karl J, Liao H, Zolg W, Guild B (2004) Quantification of C-reactive protein in the serum of patients with rheumatoid arthritis using multiple reaction monitoring mass spectrometry and ^{13}C labeled peptide standards. Proteomics 4:1175–1186
18. Gil-Dones F, Darde VM, Alonso-Orgaz S, Lopez-Almodovar LF, Mourino-Alvarez L, Padial LR, Vivanco F, Barderas MG (2012) Inside human aortic stenosis: a proteomic analysis of plasma. J Proteomics 75: 1639–1653
19. Hager JW (2002) A new linear ion trap mass spectrometer. Rapid Commun Mass Spectrom 16:512–526
20. Hopfgartner G, Varesio E, Tschappat V, Grivat C, Bourgogne E, Leuthold LA (2004) Triple quadrupole linear ion trap mass spectrometer for the analysis of small molecules and macromolecules. J Mass Spectrom 39:845–855
21. Unwin RD, Griffiths JR, Leverentz MK, Grallert A, Hagan IM, Whetton AD (2005) Multiple reaction monitoring to identify sites of protein phosphorylation with high sensitivity. Mol Cell Proteomics 4:1134–1144

Chapter 15

Proteomic Analysis of Plasma of Patients with Left Ventricular Remodeling After Myocardial Infarction: Usefulness of SELDI-TOF

Florence Pinet

Abstract

SELDI-TOF and depletion of major blood proteins is one of the most promising approaches for accessing low-abundance biomarkers. The use of combinatorial peptide ligand library (CPLL) for selecting the low-abundance proteins and of liquid-phase isoelectric focusing for purifying the corresponding proteins was tested in plasma or serum from patients with myocardial infarction (MI).

Here, we describe the SELDI profiling of CPLL-treated plasma to select low-abundance proteins in plasma and the strategy for purification and mass spectrometry identification. This approach shows the potential to select and identify candidate biomarkers in patients with left ventricular remodeling after MI.

Key words SELDI-TOF, Plasma, Protein purification, Combinatorial peptide ligand library, Liquid-phase isoelectric focusing

1 Introduction

Different proteomic techniques are available to search for novel biomarkers in heart failure and left ventricular remodeling (LVR) (1). SELDI-TOF which is a combination of chromatography on ProteinChip arrays and mass spectrometry (2, 3) offers a high-throughput technology to discover biomarkers. In the past, we used this approach to identify in plasma from patients with myocardial infarction candidate biomarkers of LVR (4). We selected and identified posttranslational variants of $\alpha 1$-chain of haptoglobin corresponding to *m/z* 9,493, 9,565, and 9,623 elevated in plasma of patients with high LVR (5).

However, this type of analysis is rendered difficult by the complexity of plasma proteome with 22 proteins accounting for 99 % of the proteome (6). The challenge is to reach the remaining 1 %, known as "deep proteome." The concentration of cardiac proteins ranges from 1 μg/mL to 1 pg/mL and is part of deep proteome (Fig. 1). A novel approach, the combinatorial peptide ligand library

Fernando Vivanco (ed.), *Vascular Proteomics: Methods and Protocols*, Methods in Molecular Biology, vol. 1000, DOI 10.1007/978-1-62703-405-0_15,

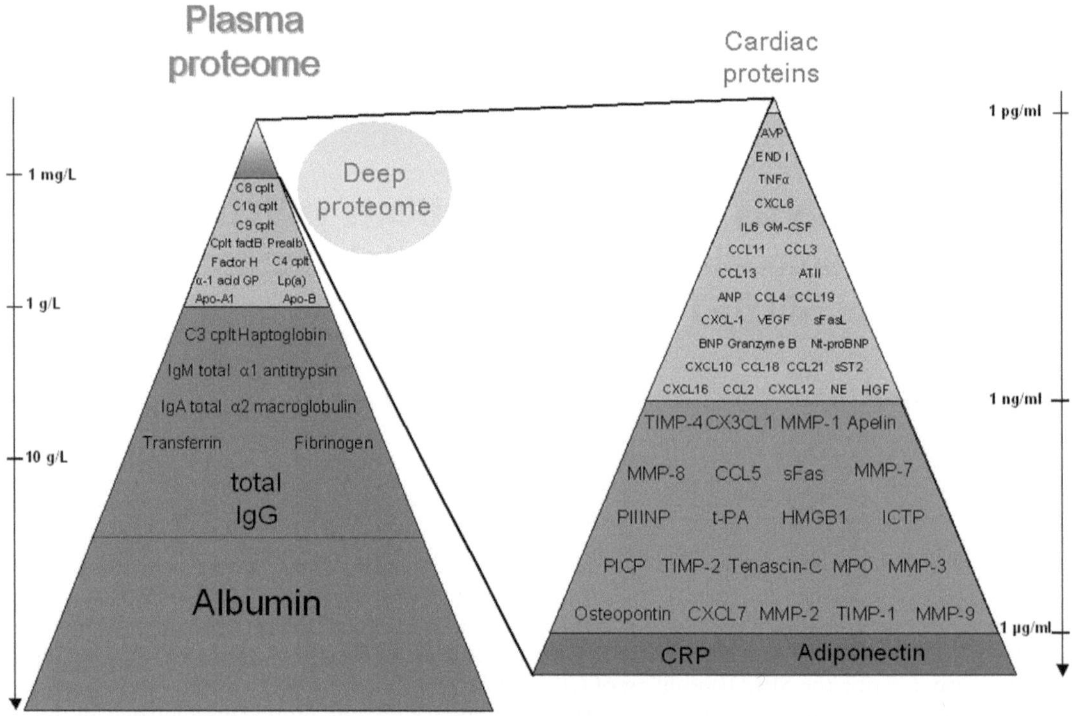

Fig. 1 Concentrations of proteins plasma. Cardiac proteins are part of deep proteome (*right triangle*), consisting of 1 % of the proteome, while 22 proteins (*left triangle*) account for up to 99 % of the total amount of proteins

(CPLL) seeks to compress the protein concentration range in plasma, through a simultaneous one-step dilution of high-abundance and concentration of low-abundance proteins (7). We have found that CPLL reproducibly increased resolution, with loss of the most abundant plasma proteins, and improved the intensity of low-abundance proteins (8, 9).

The CPLL treatment of plasma samples combined with SELDI-TOF technology has proved its usefulness to select candidate markers representing low-abundance proteins or proteolytic fragment of proteins (8). SELDI selection of differentially expressed peaks from CPLL-treated samples, however, should be followed by mass spectrometry to identify the corresponding proteins. Purification steps of SELDI peaks are critical, and an alternative technique of separation of proteins using the Microrotofor™ liquid-phase IEF cell seemed helpful (10). This approach fractionates proteins in free solution according to their isoelectric point, and this step reduces overall sample complexity and increases the concentration of low-abundance proteins relative to the original sample. We recently showed that this strategy for purification of SELDI peaks from CPLL-treated samples was effective and allowed the identification of the corresponding proteins (11).

2 Materials

Prepare all solutions using ultrapure water (deionized water with a sensitivity of 18 Ω cm at 25 °C) and analytical grade reagents. Prepare and store all reagents at room temperature (unless indicated otherwise). Follow all waste disposal regulations when disposing waste materials. We do not add sodium azide to the reagents.

2.1 Combinatorial Peptide Ligand Library Components

Combinatorial peptide ligand library (CPPL) components are in ProteoMiner Enrichment kit (Bio-Rad) (*see* **Note 1**).

1. Spin column containing 500 μL of suspended beads: 20 % beads in ethanol 20 % and acetonitrile (ACN) 0.5 %.
2. Wash buffer: 50 ml PBS buffer (150 mM NaCl, 10 mM Na_2HPO_4, pH 7.4).
3. Rehydration reagent: 5 mL of 5 % acetic acid.
4. Elution reagent: Vial of lyophilized 8 M urea and 2 % CHAPS. Reconstitute 1 vial with 610 μL of rehydration reagent (*see* **Note 2**).

2.2 SELDI-TOF Components

1. ProteinChip Arrays: Q10 (Strong Anion Exchanger), CM10 (Weak Cation Exchanger), H50 (Reverse phase) and NP20.
2. Binding buffers for ProteinChip Arrays. Q10: Tris 100 mM pH 9, CM10: sodium acetate 100 mM pH 4, H50: ACN 10 %, trifluoroacetic acid (TFA) 0.1 %, NaCl 150 mM, NP20: H_2O.
3. Washing buffer: HEPES 4 mM.
4. Lyophilized sinapinic acid (5 μg). Add 160 μL of ACN/H_2O (50/50 v/v) and 0.5 % TFA in the vial. Dilute to 50 % in the same buffer.
5. Calibrating mixture is composed of hirudin (6,964 Da), cytochrome c (12,230 Da), myoglobin (16,951 Da), carbonic anhydrase (29,023 Da), enolase (46,671 Da), albumin (64,330 Da), and IgG (147,300 Da).

2.3 Liquid-Phase Isoelectric Focusing Components

1. Liquid-phase isoelectric focusing (IEF) buffer: 7 M urea, 2 M thiourea, 4 % (v/v) CHAPS, 0.24 % Triton X-100 with glycerol (5 % v/v).
2. 2D Clean-Up kit (GE Healthcare) was used to precipitate proteins of IEF fractions before analysis.

3 Methods

Carry out all procedures at room temperature unless otherwise specified.

3.1 Sample Preparation

1. Process plasma (EDTA as anticoagulant) and serum collected within 2 h. Divide into aliquots and store at −80 °C. Do not undergo more than two freeze/thaw cycles of samples before analysis.
2. Centrifuge samples at 10,000 × *g* for 10 min to clarify.

3.2 Sample Treatment with CPLL

1. Centrifuge the spin column at 1,000 × *g* for 60 s to remove the storage solution. Discard collected material (*see* **Note 3**).
2. Add 600 μL of wash buffer to the column. Rotate column end to end several times over a 5-min period. Centrifuge the spin column at 1,000 × *g* for 60 s and discard collected material. Repeat once the process.
3. Add 1 mL of sample (plasma or serum) to the column (*see* **Note 4**). Rotate column on a rotational shaker for 2 h at room temperature (RT). Centrifuge the spin column at 1,000 × *g* for 60 s and discard collected material.
4. Add 600 μL of wash buffer to the column. Rotate column end to end several times over a 5-min period. Centrifuge the spin column at 1,000 × *g* for 60 s and discard collected material. Repeat three times more the process.
5. Add 600 μL of deionized water to the column. Rotate end to end for 1-min period. Centrifuge the spin column at 1,000 × *g* for 60 s and discard collected material.
6. Add 100 μL of elution reagent to the column. Lightly vortex for 5 s. Incubate column at RT and lightly vortex several times over a period of 15 min.
7. Place a clean collection tube (*see* **Note 5**). Centrifuge the spin column at 1,000 × *g* for 60 s. Repeat two times more the process. CPLL treatment improved the detection of low-abundance proteins and loss of the more abundant plasma protein, albumin (Fig. 2).
8. Store elution at −20 °C.

3.3 SELDI-TOF Analysis

1. Dilute tenfold 5 μL of sample (treated or not with CPPL) in binding buffer according to the arrays used.
2. Pre-equilibrate the Q10, CM10, or H50 ProteinChip arrays for 5 min in 100 μL of binding buffer before sample addition.
3. Incubate each sample (50 μL) for 30 min in binding buffer and distribute randomly in duplicate on arrays.
4. Wash each spot twice for 5 min with the binding buffer and for 5 s with HEPES buffer.

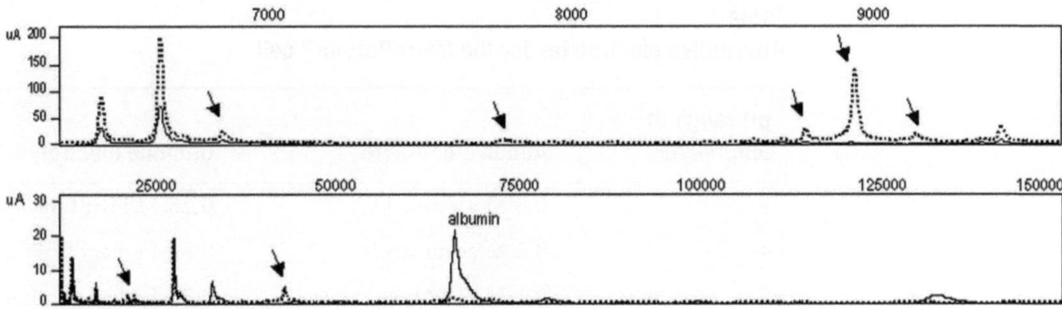

Fig. 2 Effect of plasma treatment by CPLL before SELDI-TOF analysis. Detailed spectra of low-mass proteins (6,000–10,000 Da) (*upper panel*) and of high-mass proteins (10,000–25,000 Da) (*lower panel*) of native (*dashed line*) and CPLL-treated plasma (*dotted line*). *Arrows* on spectra indicate new peaks that appeared with CPLL treatment. The peak representing albumin decreased after CPLL treatment

5. Dry partially the spots before adding 2 × 0.6 μL of sinapinic acid (*see* **Note 6**).
6. Read all arrays in an automated PBS 4000 SELDI-TOF MS. Ten laser shots per pixel were averaged. Focus mass was 16,000 Da for low mass (0–30,000 Da) and 40,000 Da for high mass (<30,000 Da) (*see* **Note 7**).
7. Process data with the ProteinChip Data Manager software. Assemble groups of peaks of similar mass into clusters, according to two-step parameter settings: (1) detection of peaks automatically according to S/N (4 or 5) and the minimum valley depth (3 or 4), in at least 10 % of all spectra, with an *m/z* error of less than 0.2 %; (2) selection of peaks according to S/N of 2 and a minimum valley depth of 2.
8. Perform univariate analysis with a nonparametric test once the peaks are clustered, and calculate the *p*-value associated with each cluster (*see* **Note 8**).

3.4 Purification of Proteins

1. Purify the proteins corresponding to the selected SELDI peaks using IEF with the MicroRotofor™ liquid-phase IEF cell (Bio-Rad Laboratories) (*see* **Note 9**).
2. After assembly, equilibrate the ion exchange membranes. Add 0.6 mL of 0.1 M H_3PO_4 buffer through the vent hole of the anode assembly and 0.6 mL of 0.1 M NaOH buffer through the vent hole of the cathode assembly.
3. Load samples (0.5 mL) into the focusing chamber of the MicroRotofor™ liquid-phase IEF cell. Fill the electrode assemblies with appropriate electrolytes depending on the pH range investigated (Table 1) and incubate overnight.
4. Perform focusing at 20 °C under a constant power of 1 W (*see* **Note 10**).

Table 1
Alternative electrolytes for the MicroRotofor® cell

pH range of ampholytes	Anode electrolyte	Cathode electrolyte
3–5	0.5 M acetic acid	0.25 M HEPES
4–6	0.5 M acetic acid	0.5 M ethanolamine
5–7	0.1 M H_3PO_4	0.5 M ethanolamine
6–8	0.1 M H_3PO_4	0.1 M NaOH
7–9	0.25 M MES	0.1 M NaOH
8–10	0.25 MES	0.1 M NaOH

5. Clean fractions for further analysis with 2D Clean-Up kit (*see* **Note 11**). Add 300 μL of precipitant and 300 μL of coprecipitant to the fraction recuperated and incubate for 15 min at RT. Centrifuge at 1,000 × *g* for 5 min. Resuspend pellet in 25 μL H_2O and add 1 mL of wash buffer and 5 μL wash additive and incubate for 30 min at −20 °C. Centrifuge at 1,000 × *g* for 5 min, dry the pellet, and resuspend in 20 μL H_2O.
6. Use NP20 arrays for controlling the purification process. Example of the SELDI peak 15,935 *m/z* purified by IEF in fraction 10 of pH gradient 3–10 (Fig. 3).

4 Notes

1. Store the unopened kit at 4 °C.
2. After reconstituting, lyophilized elution reagent should be stored at −20 °C. It is better to use freshly reconstituted elution reagent.
3. Do not discard caps.
4. Protein concentration in sample should be ≥50 mg/mL.
5. The tube contained the eluted proteins. Elution may be pooled.
6. The spots need to be completely dried before SELDI-TOF analysis.
7. Before each SELDI-TOF analysis, a calibration step is performed, using the same parameters as the sample analysis.
8. Duplicates were averaged before any statistical analysis.
9. The focusing chamber needs to be incubated overnight in SDS 20 % to keep the MicroRotofor™ liquid-phase IEF cell cleaned for the next IEF.

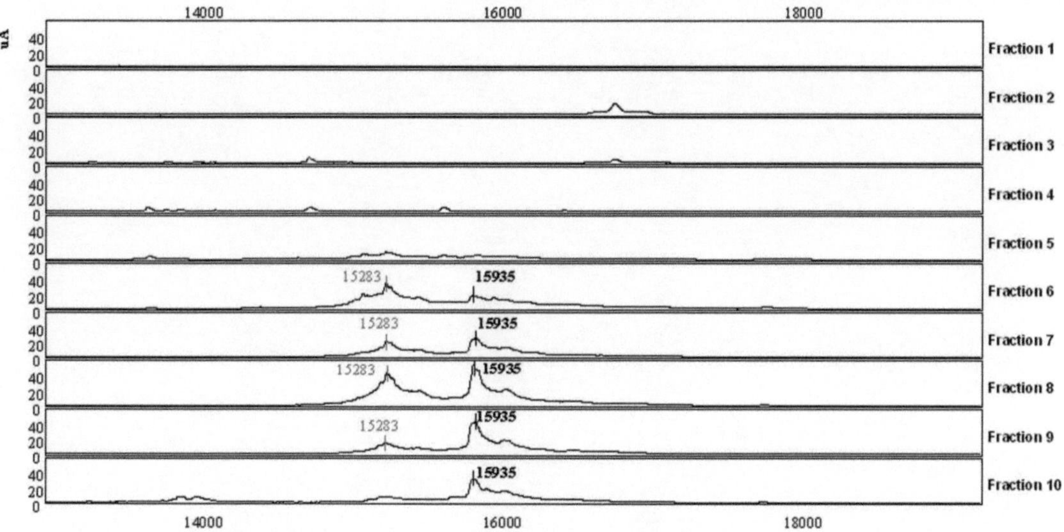

Fig. 3 Purification of SELDI peak 15,935 *m/z* using the MicroRotofor™ liquid-phase IEF cell and pH gradient 3–10. Each fraction is analyzed on NP20 array. The peak of 15,935 *m/z* is purified in fraction 10

10. At the end of IEF, protein fractions from each compartment should be harvested quickly to avoid diffusion of separated proteins.
11. Precipitation was performed to selectively discard contaminant form proteins in IEF fractions.

References

1. Dubois E, Fertin M, Burdese J et al (2011) Cardiovascular proteomics: translational studies to develop novel biomarkers in heart failure and left ventricular remodeling. Proteomics Clin Appl 5:57–66
2. Reddy G, Dalmasso EA (2003) SELDI ProteinChip® array technology: protein-based predictive medicine and drug discovery applications. J Biomed Biotechnol 2003:237–241
3. Issaq HJ, Conrads TP, Prieto DA et al (2003) SELDI-TOF MS for diagnostic proteomics. Anal Chem 75:148A–155A
4. Savoye C, Equine O, Tricot O et al (2006) Left ventricular remodeling after anterior wall acute myocardial infarction in modern clinical practice (from the Remodelage VEntriculaire (REVE) study group). Am J Cardiol 98: 1144–1149
5. Pinet F, Beseme O, Cieniewski-Bernard C et al (2008) Predicting left ventricular remodeling after a first myocardial infarction by plasma proteome analysis. Proteomics 8:1798–17808
6. Anderson NL, Polanski M, Pieper R et al (2004) The human plasma proteome: a non redundant list developed by combination of four separate sources. Mol Cell Proteomics 3:311–326
7. Righetti PG, Boschetti E, Lomas L et al (2006) Protein equalizer technology: the quest for a "democratic proteome". Proteomics 6: 3980–3992
8. Fertin M, Beseme O, Duban S et al (2010) Deep plasma proteomic analysis of patients with left ventricular remodeling after a first myocardial infarction. Proteomics Clin Appl 4:654–673
9. Beseme O, Fertin M, Drobecq H et al (2010) Combinatorial peptide ligand library plasma treatment: advantages for accessing low-abundance proteins. Electrophoresis 31: 2697–2704
10. Hey J, Posch A, Cohen A et al (2008) Fractionation of complex protein mixtures by liquid-phase isoelectric focusing. Methods Mol Biol 424:225–2239
11. Fertin M, Burdese J, Beseme O et al (2011) Strategy for purification and mass spectrometry identification of SELDI peaks corresponding to low-abundance plasma and serum proteins. J Proteomics 74:420–430

Chapter 16

Proteomic Analysis of Urinary Exosomes in Cardiovascular and Associated Kidney Diseases by Two-Dimensional Electrophoresis and LC–MS/MS

Irene Zubiri, Fernando Vivanco, and Gloria Alvarez-Llamas

Abstract

Urinary exosomes are membranous vesicles 40–100 nm in size containing proteins that are characteristic of every renal tubule epithelial cell type. In this chapter, we describe a methodology to isolate and analyze urinary exosomes proteome by 2-DE and LC–MS/MS, in the search for biomarkers of vascular and associated kidney diseases. We describe an isolation methodology by serial (ultra)centrifugation steps compatible with 2-DE and LC–MS/MS analysis. Exosome purity is confirmed by electron microscopy and Western blot.

Key words Exosomes, Urine, 2-DE, LC–MS/MS, Vascular disease, Kidney diseases

1 Introduction

Exosomes are membranous vesicles 40–100 nm in size, released by cells in different extracellular fluids. Blood exosomes, despite being previously considered inert debris without specific function, have been recently demonstrated to be important in the regulation of different mechanisms in vascular pathologies (1). Urinary exosomes contain proteins that are characteristic of every renal tubule epithelial cell type and from the urinary collecting system. They can shuttle information between nonrenal cells via transfer of protein and RNA, providing relevant information to a variety of disease processes and constituting a promising source of biomarkers (2) The proteome of urinary exosomes has been studied in the search for candidate biomarkers of kidney diseases (2–7) and a variety of nonrenal disorders as lung cancer (8), hepatic disease (9), prostate (10–12) and bladder cancer (13), among others.

Exosomes represent only a 3 % of the whole urine proteome (14). Its isolation and characterization is extremely challenging but otherwise constitutes an enriched subproteome with reduced complexity compared to the urinary proteome. In this way, low-abundance

Fernando Vivanco (ed.), *Vascular Proteomics: Methods and Protocols*, Methods in Molecular Biology, vol. 1000,
DOI 10.1007/978-1-62703-405-0_16,

proteins that may have pathophysiological significance can be enriched against high-abundance proteins.

In this chapter, we describe a methodology to isolate and analyze urinary exosomes proteome by 2-DE and LC–MS/MS, in the search for biomarkers of vascular and associated kidney diseases. The Tamm–Horsfall protein (THP) also called uromodulin is the most abundant soluble protein in normal urine, and its polymerization can entrap urinary exosomes, reducing the isolation reproducibility and exosome recovery. To avoid this deleterious effect and potential interference with downstream applications, THP can be depolymerized with reducing agents and heat (15). Albumin can also represent a problem when studying diseases that lead to a high content in albumin and IgG light and heavy chain in urine, and this condition can affect the extraction causing a contamination in the exosome pellet.

We describe here an isolation methodology by serial (ultra) centrifugation steps after which exosomes purity is confirmed by electron microscopy and Western blot. Depletion of three major proteins present in the exosome fraction: THP, IgG, and albumin (high-abundance urine protein in chronic kidney disease patients), is also contemplated, which otherwise could mask detection of minor proteins candidate for biomarkers.

2 Materials

2.1 Urine Collection

1. Protease inhibitor cocktail containing 104 mM 4-(2-aminoethyl) benzenesulfonyl fluoride hydrochloride (AEBSF), 80 μl aprotinin, 2 mM leupeptin, 1.5 mM pepstatin, and 1.4 mM E-64 (Sigma P8340).
2. Sterile plastic recipients.

2.2 Exosome Isolation

1. Isolation solution: 10 mM triethanolamine and 250 mM sucrose in double-distilled water (ddH_2O). Adjust pH to 7.6.
2. Centrifuge tubes polypropylene (50 ml, ref: 357005 Beckman).
3. Polycarbonate Aluminum Bottle with Cap Assembly (26.3 ml 355618.Beckman).
4. JA 20 Rotor-fixed angle (Beckman).
5. 70 Ti Rotor-fixed angle (Beckman).

2.3 Depletion of Urine Major Proteins

1. THP depletion: 200 mg/ml dithiothreitol (DTT, Bio-Rad) in isolation solution.
2. Albumin and IgG depletion: ProteoPrep® Immunoaffinity Albumin and IgG Depletion Kit (Sigma).

2.4 Exosome Characterization

1. Electron microscopy:
   (a) 4 % Paraformaldehyde in PBS-T.
   (b) 1 % Uranyl acetate in ddH_2O.
   (c) Parafilm.
   (d) Formvar–carbon-coated EM grids (Ted Pella).
2. Western blot:
   (a) Acrylamide/bisacrylamide (ProtoGel), 1 M Tris–HCl pH8.8, sodium dodecyl sulfate (SDS, Bio-Rad) (10 % solution), Tetramethyl-ethylenediamine (TEMED, Bio-Rad), ammonium persulfate (APS, Bio-Rad) (10 % solution, freshly prepared) for polyacrylamide running gels of desired size and composition.
   (b) Acrylamide/bisacrylamide, 1 M Tris–HCl pH6.8, SDS (10 % solution), TEMED, APS (10 % solution, freshly prepared) for the stacking gel of desired size and composition.
   (c) Running buffer 1×: 0.03 M Tris, 0.2 M glycine, 0.1 % SDS.
   (d) Sample buffer (4×): 252 mM Tris–HCl, 40 % Glycerol, 8 % SDS. Store at room temperature and warm to redissolve and add bromophenol blue and 5 % β-mercaptoethanol just before use.
   (e) Nitrocellulose membranes (Bio-Rad).
   (f) Blocking and antibody solutions: 5 % milk in PBS-T.
   (g) Exosome membrane marker antibodies: Alix, 1:500 (Santa Cruz), Tsg101, 1:500 (Abcam) and negative marker antibody Calnexin, 1:2,500 (Enzo life sciences), diluted in 5 % milk in PBS-T.
   (h) HRP-conjugated secondary antibodies: rabbit anti-mouse 1:2,500, goat anti-rabbit 1:5,000 (Santa Cruz), diluted in 5 % milk in PBS-T.

2.5 Two-Dimensional Polyacrylamide Gel Electrophoresis (2D-PAGE)

1. Sample solubilization, purification, and quantification:
   (a) Lysis buffer: 7 M urea, 2 M thiourea, 4 % 3-((3-cholamidopropyl) dimethylammonio)-1-propanesulfonate (CHAPS) in ddH_2O. Store at −20 °C.
   (b) 2D Clean-up kit (GE Healthcare).
   (c) BSA (bovine serum albumin, Sigma) in ddH_2O.
   (d) Bradford ultra (Expedeon).
2. Isoelectricfocusing:
   (a) Immobiline DryStrip pH 3–11, 18 cm (GE Healthcare).
   (b) Mineral oil (Bio-Rad).

(c) Rehydration buffer: 0.8 % Ampholytes (IPG Buffer pH 3–11 NL GE Healthcare), 10 mM DTT, and traces of bromophenol blue in lysis buffer just before use.

(d) IPGphor (GE Healthcare).

3. Equilibration:

Equilibration buffer: 1 % DTT for the first step or 2,5 % iodoacetamide (IAA) for the second step in 75 mM Tris, 6 M urea (Bio-Rad), 40 % Glycerol, 2 % SDS, and bromophenol blue (Bio-Rad) traces in ddH_2O. Store at −20 °C.

4. SDS-PAGE:

(a) Prepare polyacrylamide gels of desired size and composition as described above (Subheading 2.4, **step 2** (a, b)).

(b) Running buffer (1×): described above (Subheading 2.4, **step 2**(c)).

(c) Ettan DALTsix (GE Healthcare).

(d) Silver staining kit, protein PlusOne (GE Healthcare), prepare different solutions as described here:

- Fixative: 40 % ethanol, 10 % acetic acid.
- Sensitizer: 30 % ethanol, 4 % sodium thiosulfate (5 % w/v solution), 68 mg/ml sodium acetate.
- Silver solution: 0.25 % silver nitrate.
- Developer: 25 mg/ml sodium carbonate with 20–40 μl formaldehyde (37 % w/v solution) per 100 ml total volume.
- Stopper: 14.6 mg/ml EDTA.

2.6 Protein Digestion and Peptide Analysis by nLC–MS/MS Analysis

1. Tryptic digestion: DTT (Bio-Rad), IAA (Bio-Rad), modified porcine trypsin (Promega), trifluoroacetic acid, acetic acid, formic acid (Merck).

2. nLC–MS/MS:

- Mobile phase (A) 0.5 % acetic acid or 0.1 % formic acid.
- Mobile phase (B) 0.1 % formic acid in 100 % acetonitrile.
- C-18 reversed-phase nano-column (100 mm i.d. and 12 cm, Mediterranean Sea, Teknokroma).
- BioBasic C-18 PicoFrit column (75 mm, 10 cm; New Objective).
- LTQ Orbitrap XL mass spectrometer (Thermo Fisher, San José, CA, USA) and LTQ linear ion trap mass spectrometer (Thermo Electron, San Jose, CA).
- Data analysis: Scaffold software (Proteome Software Inc., Portland, OR).

3 Methods

3.1 Urine Collection

1. Collect the second morning urine in a sterile recipient (100 ml).
2. Add protease inhibitors, essential for exosome preservation; 50 μl/100 ml.
3. Maintain at room temperature and process immediately or store at −80 °C (*see* **Note 1**).

3.2 Exosome Isolation

1. Low-speed centrifugation: 17,000 × *g* for 10 min at 4 °C, for eliminating cell debris and membrane fractions (*see* **Note 2**).
2. Take the supernatant to ultracentrifuge bottles and centrifuge at 200,000 × *g* for 70 min at 4 °C (*see* **Note 3**).
3. Discard supernatant and resuspend the exosome pellet in the appropriate buffer, depending on the next step as follows:
   (a) To perform a depletion step, solubilized pellet in 50 μl of isolation solution (*see* **Notes 4** and **5**).
   (b) To further purify the exosome pellet, solubilize in 50 μl of isolation solution, preferably transfer to a clean tube; add 20 ml of isolation solution; and repeat the ultracentrifugation step. Discard the supernatant and solubilize the pellet in 50 μl lysis buffer (*see* **Notes 6** and 7).
   (c) If neither cleaning nor depletion procedure is desirable, solubilize the pellet in 50 μl lysis buffer.

3.3 THP and Albumin Depletion (Optional Step)

1. Tamm–Horsfall protein depletion:

   Once solubilized the exosome pellet in 50 μl isolation solution, add 200 mg/ml DTT to reduce THP protein and avoid aggregation. Heat at 95 °C for 2 min. Add isolation solution up to 20 ml and repeat the ultracentrifugation step. Discard the supernatant and resuspend the pellet in 50 μl lysis buffer.
2. Albumin and IgG depletion:

   Use ProteoPrep® Immunoaffinity Albumin and IgG Depletion commercial kit following the commercial protocol (*see* **Note 8**). Transfer the eluted fraction to a clean ultracentrifuge tube, add 20 ml of isolation solution, and repeat the ultracentrifugation step. Discard supernatant and resuspend the pellet in 50 μl lysis buffer.

3.4 Electron Microscopy and Western Blotting (Figs. 1 and 2)

1. Electron microscopy:
   (a) Resuspend exosome pellet in 4 % paraformaldehyde in PBS pH 7.2.
   (b) Spot 15 μl sample on parafilm.
   (c) Float a Formvar–carbon-coated grid on sample droplet for 10 min.

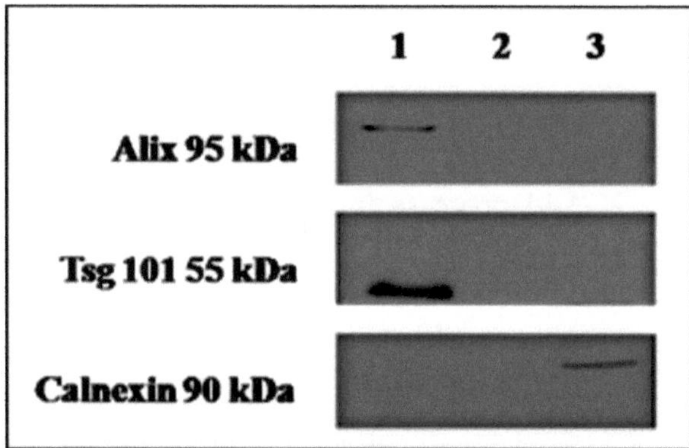

Fig. 1 Alix and Tsg101 (specific exosome membrane markers) have been detected in exosomes (*lane 1*) but neither in urine (*lane 2*) nor in tubular epithelial cell line HK2 (*lane 3*). Calnexin (specific membrane marker of rough endoplasmic reticulum) has been detected in HK2 cell lysate but neither in exosomes nor in urine

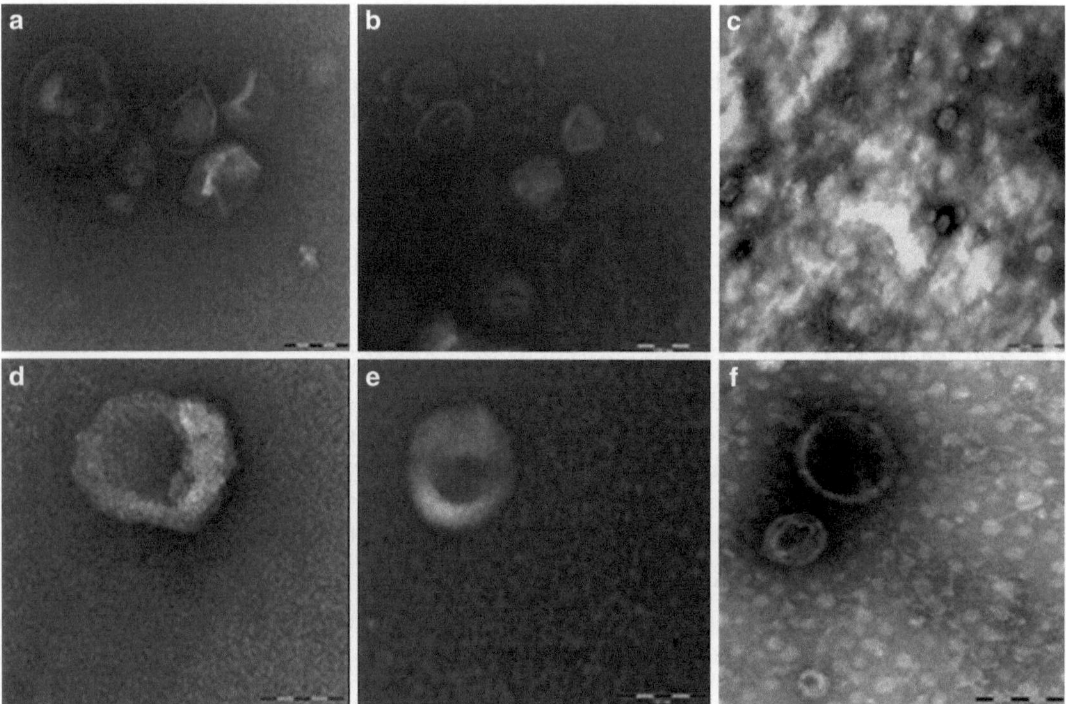

Fig. 2 (**a**–**e**) Electronic microscopy images of urinary exosomes after THP depletion, **c** and **f**, exosomes in the presence of THP

(d) Use a filter paper to eliminate the excess fluid from the grid.

(e) Float the grid on 15 μl ddH_2O water for 5 min.

(f) Negative stain by floating grid on 20 μl of 2 % uranyl acetate in ddH_2O for 30 s.

   (g) Use a filter paper to eliminate the excess fluid from the grid.
   (h) Let grid dry on filter paper (coated side up) for 5 min to dry out.

2. Western blot:
   (a) Solubilize exosome pellet in lysis buffer and quantify protein content by Bradford assay.
   (b) Using the appropriate volume for 25 μg of exosomal proteins, add sample buffer 4× to a final concentration of 1×. Add 5 % β-mercaptoethanol and bromophenol blue traces. Heat 5 min at 95 °C.
   (c) Load samples in a 10 % acrylamide gel (with 4 % stacking gel) for performing the electrophoresis.
   (d) Transfer the proteins from the gel to a nitrocellulose membrane by electroblotting 10 V for 1 h.
   (e) Block the membrane using 5 % milk in PBS-T for 1 h at room temperature.
   (f) Incubate the membrane overnight at 4 °C with specific antibodies.
   (g) Wash the membrane for 10 min in PBS-T for a total of five times.
   (h) Incubate the membranes with HRP-conjugated secondary antibody for 1 h.
   (i) Wash the membrane for 10 min in PBS-T for a total of five times.

3.5 Two-Dimensional Electrophoresis (Fig. 3)

1. Sample cleaning and quantification:
   (a) Clean the exosome sample solubilized in 50 μl lysis buffer using 2D Clean-up kit following manufacturer's protocol.
   (b) Resuspend the pellet in 30 μl lysis buffer and quantify it by Bradford assay.

2. Isoelectricfocusing:
   (a) Add 10 mM DTT and 0.8 % Ampholytes to the sample and traces of bromophenol blue.
   (b) Load samples (containing, i.e., 80–150 μg total protein), using the cup-loading approach onto IPG strips previously rehydrated with rehydration buffer. Add mineral oil to cover cups and strips.
   (c) Perform IEF using the following program: 200 V for 1 h, 500 V 1 h, 1,000 V for 1 h, from 1,000 V to 8,000 V in 1 h, and 8,000 V until 60,000 V are accumulated.

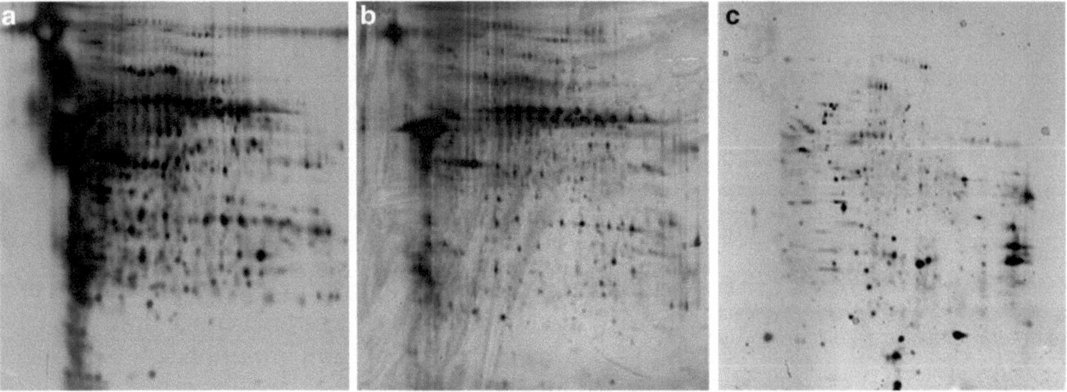

Fig. 3 Images of 2-DE silver-stained gels corresponding to isolated urinary exosomes without major proteins depletion (**a**), after THP depletion (**b**), and after THP, albumin, and IgG depletion (**c**)

3. Equilibration:
   (a) Equilibrate each strip with 5 ml 1 % DTT equilibration buffer for 15 min. Discard and perform the second equilibration step with 5 ml 2 % IAA equilibration buffer.
   (b) Rinse the strip with running buffer.
4. SDS-PAGE:

   Perform second dimension on 10 % polyacrylamide gels using Ettan DALTsix System (GE Healthcare) (*see* **Note 9**).
5. Silver staining of 2-DE gels:
   (a) Add fixing solution (250 ml/gel), 30 min at least (*see* **Note 10**).
   (b) Discard fixing solution and add sensitizer solution for 20 min (*see* **Note 11**).
   (c) Discard sensitizer solution, wash the gel in ddH_2O for 5 min 250 ml/gel; repeat this step twice for a total of three wash cycles.
   (d) Add silver solution (250 ml/gel) for 15 min.
   (e) Discard silver solution, wash the gel in ddH_2O for 1 min, and repeat this step (*see* **Note 12**).
   (f) Discard ddH_2O solution and add developer solution 250 ml/gel.
   (g) Discard developer solution and add Stopper solution 250 ml/gel for 10 min.
   (h) Discard stopper solution, wash the gel in ddH_2O for 10 min 250 ml/gel, and repeat this step.

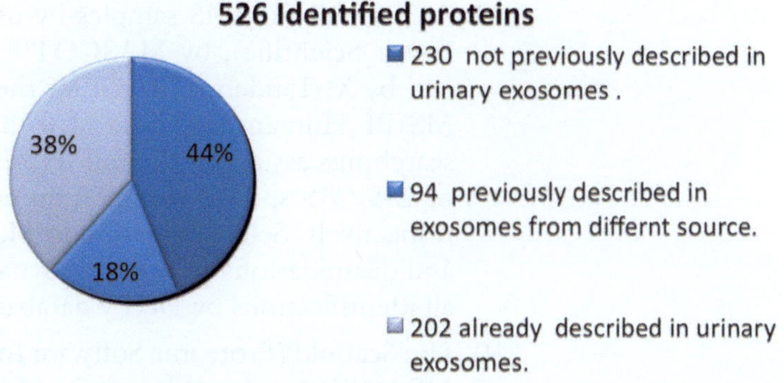

Fig. 4 LC–MS/MS analysis of (THP depleted) urinary exosomes (healthy control) performed with an Orbitrap™ instrument

3.6 LC–MS/MS Label Free Analysis (Fig. 4)

1. Clean and concentrate the sample:
   (a) Load the sample in a SDS-PAGE and run the electrophoresis until the sample reaches the running gel.
   (b) Stop the electrophoresis and stain the gel in Coomassie blue solution.
   (c) Excise the unique gel band where all proteins are concentrated.
2. Reduce with10mM DTT and alkylate cysteine residues with 50 mM IAA.
3. Digest the sample overnight at 37 °C with modified porcine trypsin (added at a final ratio of 1:50 trypsin–protein) in 50 mM ammonium bicarbonate.
4. Stop reaction with trifluoroacetic acid and vacuum-dry samples.
5. Dissolve them in 1 % acetic acid for LC–MS analysis.
6. Analyze the resulting tryptic peptide mixtures onto a C-18 reversed-phase nano-column in a continuous acetonitrile gradient consisting of 0–43 % B in 90 min, 50–90 % B in 1 min (B=90 % acetonitrile, 0.5 % acetic acid). Use a flow rate of 300 nL/min to elute peptides from the RP nano-column to an emitter nanospray needle for real-time ionization and peptide fragmentation on an LTQ Orbitrap mass spectrometer.
7. Analyze the most intense ten parent ions along the chromatographic run (130 min) by an enhanced FT-resolution spectrum (resolution = 30,000) followed by the MS/MS spectra. Set dynamic exclusion at 30 s.
8. For protein identification, extract tandem mass spectra and charge state deconvolute using Proteome Discoverer 1.2.0.207 (Thermo Fisher Scientific).

9. Analyze all MS/MS samples by using SEQUEST™ (Thermo Fisher Scientific), by MASCOT™ program (Matrix Science), and by X! Tandem (The GPM, thegpm.org). Set up to search MSIPI_Human_3.67.fasta (1.0, 87,040 entries). Perform all searchings assuming the full trypsin digestion. Set for full MS or MS/MS spectra searches, an error of 15 ppm or 0.8 Da, respectively. Select oxidation in M, phosphorylation in S or T, and deamidation in Q or N as dynamic modifications. Perform all identifications by Decoy database search for FDR analysis.
10. Use Scaffold (Proteome Software Inc., Portland, OR) to analyze MS/MS peptide and protein identifications. Assign protein probabilities by the ProteinProphet algorithm. Proteins that contain similar peptides and cannot be differentiated based on MS/MS analysis alone will be grouped to satisfy the principles of parsimony.

4 Notes

1. In case of freezing, extensive vortexing after thawing maximizes the recovery of urinary exosomes. Urinary exosomes remain intact during long-term storage (16).
2. In this low-speed spin pellet, together with cell membranes and debris, there are THP large polymers which constitute a network which may trap exosomes into this pellet. By treatment with a reducing agent, the THP network is disaggregated and exosomes are released. In case this treatment is performed, add 10 ml of isolation solution, and repeat the low-speed centrifugation (17,000 × *g*), recovering the supernatant and pooling it together with the first 17,000 × *g* centrifuge supernatant. This extra step increases exosomal recovery from urine and improves reproducibility isolation (17).
3. 4°C will be the appropriate temperature only in case the samples had been previously frozen. As THP tends to aggregate at cold temperature, if the protocol is being used with fresh samples (not frozen) 17,000 × *g* and 200,000 × *g*, centrifugation steps have to be performed at 25 °C to avoid this polymerization.
4. 50 μl is usually enough to fully solubilized the exosome pellet. If not, a higher volume can be used. No significant differences are appreciated up to 500 μl buffer in terms of quality of extraction, although following steps could be more difficult to handle.
5. The same 50 μl will be used to solubilize the pellet from different tubes. The first pellet will be resuspended in this volume by pipetting up and down and will be transferred to the next tube. Minimize the number of tubes used in the ultracentrifugation step to increase exosome recovery (solubilization is harder when more tubes are used). Find a reasonable number of tubes

for a good exosome recovery, avoiding the protocol to become very time consuming (15).

6. The volume can be varied depending on the tubes used, but it is essential to add a minimum volume to avoid bottles collapse and breakage into the rotor (i.e., 20 ml capacity bottles need to be filled at least 17 ml).
7. Exosome recovery will be less efficient when depleting THP and albumin or further cleaning the sample. Performing an extra purification step usually reduces the efficiency of a process.
8. Start with an initial sample volume of 50–70 μl (depending on the volume used to solubilize the exosome pellet) instead of 40 μl as detailed in the manufacturer's protocol.
9. Upper chamber can be filled with running buffer SDS 2× (0.02 % SDS instead of 0, 01 % SDS)
10. Prepare the fixing solution, just before use. Pause point: gels can be maintained in fixing solution.
11. Prepare the sensitizer solution in advance and store it at 4 °C. It is recommendable to use it cold.
12. It is important not to wash the gels longer than 1 min.

Acknowledgments

This work was supported by grants from FIS CP09/229, FIS PI11/01401, and FIS PI08/0970.

References

1. Azevedo LC, Pedro MA, Laurindo FR (2007) Circulating microparticles as therapeutic targets in cardiovascular diseases. Recent Pat Cardiovasc Drug Discov 2:41–51
2. Street JM, Birkhoff W, Menzies RI, Webb DJ, Bailey MA, Dear JW (2011) Exosomal transmission of functional aquaporin 2 in kidney cortical collecting duct cells. J Physiol 589: 6119–6127
3. Rood IM, Deegens JK, Merchant ML, Tamboer WP, Wilkey DW, Wetzels JF, Klein JB (2010) Comparison of three methods for isolation of urinary microvesicles to identify biomarkers of nephrotic syndrome. Kidney Int 78:810–816
4. Sonoda H, Yokota-Ikeda N, Oshikawa S, Kanno Y, Yoshinaga K, Uchida K, Ueda Y et al (2009) Decreased abundance of urinary exosomal aquaporin-1 in renal ischemia-reperfusion injury. Am J Physiol Renal Physiol 297: F1006–F1016
5. Goligorsky MS, Addabbo F, O'Riordan E (2007) Diagnostic potential of urine proteome: a broken mirror of renal diseases. J Am Soc Nephrol 18:2233–2239
6. van Balkom BW, Pisitkun T, Verhaar MC, Knepper MA (2011) Exosomes and the kidney: prospects for diagnosis and therapy of renal diseases. Kidney Int 80:1138–1145
7. Zhou H, Cheruvanky A, Hu X, Matsumoto T, Hiramatsu N, Cho ME, Berger A et al (2008) Urinary exosomal transcription factors, a new class of biomarkers for renal disease. Kidney Int 74:613–621
8. Arscott WT, Camphausen KA (2011) Analysis of urinary exosomes to identify new markers of non-small-cell lung cancer. Biomark Med 5:822
9. Conde-Vancells J, Rodriguez-Suarez E, Gonzalez E, Berisa A, Gil D, Embade N et al (2010) Candidate biomarkers in exosome-like vesicles

purified from rat and mouse urine samples. Proteomics Clin Appl 4:416–425

10. Mitchell PJ, Welton J, Staffurth J, Court J, Mason MD, Tabi Z, Clayton A (2009) Can urinary exosomes act as treatment response markers in prostate cancer? J Transl Med 7:4
11. Duijvesz D, Luider T, Bangma CH, Jenster G (2011) Exosomes as biomarker treasure chests for prostate cancer. Eur Urol 59:823–831
12. Lu Q, Zhang J, Allison R, Gay H, Yang WX, Bhowmick NA, Frelix G, Shappel l S, Chen YH (2009) Identification of extracellular delta-catenin accumulation for prostate cancer detection. Prostate 69:411–418
13. Welton JL, Khanna S, Giles PJ, Brennan P, Brewis IA, Staffurth J, Mason MD, Clayton A (2010) Proteomics analysis of bladder cancer exosomes. Mol Cell Proteomics 9:1324–1338
14. Barratt J, Topham P (2007) Urine proteomics: the present and future of measuring urinary protein components in disease. CMAJ 177: 361–368
15. Gonzales PA, Zhou H, Pisitkun T, Wang NS, Star RA, Knepper MA, Yuen PS (2010) Isolation and purification of exosomes in urine. Methods Mol Biol 64:89–99
16. Zhou H, Yuen PS, Pisitkun T, Gonzales PA, Yasuda H, Dear JW, Gross P, Knepper MA, Star RA (2006) Collection, storage, preservation, and normalization of human urinary exosomes for biomarker discovery. Kidney Int 69: 1471–1476
17. Fernandez-Llama P, Khositseth S, Gonzales PA, Star RA, Pisitkun T, Knepper MA (2010) Tamm-Horsfall protein and urinary exosome isolation. Kidney Int 77:736–742

Index

Fernando Vivanco (ed.), *Vascular Proteomics: Methods and Protocols*, Methods in Molecular Biology, vol. 1000, DOI 10.1007/978-1-62703-405-0, © Springer Science+Business Media New York 2013

C

J

K

L

M

N

Q

R

S

MIX
Papier aus verantwortungsvollen Quellen
Paper from responsible sources
FSC® C105338

If you have any concerns about our products,
you can contact us on
ProductSafety@springernature.com

In case Publisher is established outside the EU,
the EU authorized representative is:
Springer Nature Customer Service Center GmbH
Europaplatz 3, 69115 Heidelberg, Germany

Printed by Libri Plureos GmbH
in Hamburg, Germany